Introduction

The aim of the *Primary Mathematics* curriculum is to allow students to develop their ability in mathematical problem solving. This includes using and applying mathematics in practical, real-life situations as well as within the discipline of mathematics itself. Therefore the curriculum covers a wide range of situations from routine problems to problems in unfamiliar contexts to open-ended investigations that make use of relevant mathematical concepts.

An important feature of learning mathematics with this curriculum is the use of a concrete introduction to the concept, followed by a pictorial representation, followed by the abstract symbols. The textbook does supply some concrete introductory situations, but you, as the teacher, should supply a more concrete introduction when applicable. The textbook then supplies the pictorial and abstract aspects of this progression. For some students a concrete illustration is more important than for other students.

This guide includes the following :

♦ **Scheme of Work**: A table with a suggested weekly schedule, the primary objective for each lesson, and corresponding pages from the textbook, workbook, and guide.

♦ **Manipulatives**: A list of manipulatives used in this guide.

♦ **Objectives**: A list of objectives for each chapter.

♦ **Vocabulary**: A list of new mathematical terms for each chapter.

♦ **Notes**: An explanation of what students learned in earlier levels, the concepts that will be covered in the chapter, and how these concepts fit in with the program as a whole.

♦ **Material**: A list of suggested manipulatives that can be used in presenting the concepts in each chapter.

♦ **Activity**: Teaching activities to introduce a concept concretely or to follow up on a concept in order to clarify or extend it so that students will be more successful with independent practice.

♦ **Discussion**: The opening pages of the chapter and tasks in the textbook that should be discussed with the student. A scripted discussion is not provided. You should follow the material in the textbook. Additional pertinent points that could be included in the discussion are given in this guide.

♦ **Practice**: Tasks in the textbooks students can do as guided practice or as an assessment to see if they understand the concepts covered in the teaching activity or the discussions.

♦ **Workbook**: Workbook exercises that should be done after the lesson.

♦ **Reinforcement**: Additional activities that can be used if your student needs more practice or reinforcement of the concepts. This includes referenc   the exercises in the optional *Primary Mathematics Extra Practice* book.

♦ **Games**: Optional simple games that can be used to p

- **Enrichment**: Optional activities that can be used to further explore the concepts or to provide some extra challenge.

- **Tests:** References to the appropriate tests in the optional *Primary Mathematics Tests* book.

- **Answers:** Answers to all the textbook tasks and workbook problems, and many fully worked solutions. Answers to textbook tasks are provided within the lesson. Answers to workbook exercises for the chapter are located at the end of the chapters in the guide.

- **Mental Math**: Problems for more practice with mental math strategies.

- **Appendix**: Pages containing drawings and charts that can be copied and used in the lessons.

The textbook and workbook both contain a review for every unit. You can use these in any way beneficial to your student. For students who benefit from a more continuous review, you can assign three problems or so a day from one of the practices or reviews. Or, you can use the reviews to assess any misunderstanding before administering a test. The reviews, particularly in the textbook, do sometimes carry the concepts a little farther. They are cumulative, and so allow you to refresh your student's memory or understanding on a topic that was covered earlier in the year.

In addition, there are supplemental books for *Extra Practice* and *Tests*. In the test book, there are two tests for each chapter. The second test is multiple choice. There is also a set of two cumulative tests at the end of each unit. You do not need to use both tests. If you use only one test, you can save the other for review or practice later on. You can even use the reviews in the workbook for assessment and not get the test book at all. So there are plenty of choices for assessment, review, and practice.

The mental math exercises that go along with a particular chapter or lesson are listed as reinforcement or enrichment in the lesson. They can be used in a variety of ways. You do not need to use all the mental math exercises listed for a lesson on the day of the lesson. You can have your student do one mental math exercise a day or every few days, repeating some of them, at the start of the lesson or as part of the independent work. You can do them orally, or have your student fill in the blanks, depending on the type of problems. You can have your student do a 1-minute "sprint" at the start of each lesson using one mental math exercise for several days to see if he or she can get more of the problems done each successive day. You can use the mental math exercises as a guide for creating additional "drill" exercises.

The "Scheme of Work" on the next few pages is a suggested weekly schedule to help you keep on track for finishing the textbook in about 18 weeks. No one schedule or curriculum can meet the needs of all students equally. For some chapters, your student may be able to do the work more quickly, and for others more slowly. Take the time your student needs on each topic and each lesson. For students with a good mathematical background, each lesson in this guide will probably take a day. For others, some lessons which include a review of previously covered concepts may take more than a day. There are a few lessons that are only review that your student may not need.

Use the reinforcement or enrichment activities at your discretion and according to your student's needs.

This printing of this guide was written when the latest printing of the textbook, workbook, *Extra Practice*, and *Tests* were in 2008. New printings of these books may have corrected errors or may have slight changes, which will be incorporated into later printings of this guide.

Scheme of Work

Textbook: *Primary Mathematics Textbook* 4B, Standards Edition
Workbook: *Primary Mathematics Workbook* 4B, Standards Edition
Guide: *Primary Mathematics Home Instructor's Guide* 4B, Standards Edition (this book)
Extra Practice: *Primary Mathematics Extra Practice* 4, Standards Edition
Tests: *Primary Mathematics Tests* 4B, Standards Edition

Week		Objectives	Text book	Work book	Guide
Unit 6: Decimals					
		Chapter 1: Tenths			1-2
1	1	♦ Understand the tenths place in decimal numbers. ♦ Express a proper fraction with a denominator of 10 as a decimal number. ♦ Relate 1-place decimal numbers to measures.	8-10	7-9	3-4
	2	♦ Read and write 1-place decimal numbers greater than 1. ♦ Locate 1-place decimal numbers on a number line.	11	10-11	5
	3	♦ Express a 1-place decimal number as a mixed number or fraction in simplest form. ♦ Compare and order 1-place decimal numbers. ♦ Interpret linear progressions involving 1-place decimals.	12-13	12-13	6-7
	4	♦ Rename 10 tenths as 1 one and 1 one as 10 tenths.	13	14-15	8
		Extra Practice, Unit 6, Exercise 1, pp. 99-100			
		Tests, Unit 6, 1A and 1B, pp. 1-4			
		Answers to Workbook Exercises 1-4			9
		Chapter 2: Hundredths			10-11
2	1	♦ Read and write decimal numbers for hundredths. ♦ Relate each digit in a 2-place decimal number to its place value.	14-16	16-18	12-13
	2	♦ Express a fraction as a decimal number.	17-18	19-20	14
	3	♦ Interpret and complete linear progressions involving 2-place decimals. ♦ Locate 2-place decimals on a number line.	18-19	21-22	15
	4	♦ Express a decimal number as a fraction in simplest form.	19-20	23-24	16
	5	♦ Compare and order decimal numbers.	21-22	25-26	17

Week		Objectives	Text book	Work book	Guide
3	6	♦ Use mental math addition or subtraction strategies with decimal numbers.	22	27-28	18
		Extra Practice, Unit 6, Exercise 2, pp. 101-104			
		Tests, Unit 6, 2A and 2B, pp. 5-8			
		Answers to Workbook Exercises 5-10			19
		Chapter 3: Thousandths			20
	1	♦ Read and write decimal numbers for thousandths. ♦ Interpret 3-place decimals in terms of place value. ♦ Count up or down by a hundredth or a thousandth.	23-24	29-30	21-22
	2	♦ Compare and order decimal numbers.	24-25	31	23
	3	♦ Write a decimal number as a fraction. ♦ Compare and order a mixture of whole numbers, decimals, and fractions.	25	32-33	24
		Extra Practice, Unit 6, Exercise 3, pp. 105-108			
4	4	♦ Practice.	26-27		25
		Tests, Unit 6, 3A, pp. 9-10			
		Answers to Workbook Exercises 11-13			26
		Chapter 4: Rounding			27
	1	♦ Round decimal numbers to the nearest whole number.	28-30	34-35	28
	2	♦ Round decimal numbers to the nearest tenth.	30	36	29
		Extra Practice, Unit 6, Exercise 4, pp. 109-110			
		Tests, Unit 6, 4A and 4B, pp. 11-13			
		Review 6	31-34	37-41	30
		Tests, Units 1-6, Cumulative A and B, pp. 15-22			
		Answers to Workbook Exercises 14-15			31
		Answers to Workbook Review 6			31
Unit 7: The Four Operations on Decimals					
		Chapter 1: Addition and Subtraction			32-33
5	1	♦ Add tenths to tenths and hundredths to hundredths.	35-37	42	34
	2	♦ Add 1-place decimal numbers.	37	43	35
	3	♦ Add 2-place decimal numbers.	38-39	44-45	36
	4	♦ Estimate sums.	39	46	37

Week		Objectives	Text book	Work book	Guide
	5	♦ Subtract tenths.	40	47	38
6	6	♦ Subtract hundredths.	41-42	48-49	39
	7	♦ Subtract 1-place decimal numbers.	42	50	40
	8	♦ Subtract 2-place decimal numbers.	43	51-52	41
	9	♦ Estimate sums and differences. ♦ Use mental math to add decimal numbers close to a whole.	44	53	42
	10	♦ Solve word problems.	45-46	54-56	43
7	11	♦ Practice.	47-48		44
		Extra Practice, Unit 7, Exercise 1, pp. 115-120			
		Tests, Unit 7, 1A and 1B, pp. 23-26			
		Answers to Workbook Exercises 1-10			45
		Chapter 2: Multiplication			46
	1	♦ Multiply tenths and hundredths by ones.	49-51	57-58	47
	2	♦ Multiply 1-place decimals by ones. ♦ Estimate products.	52	59	48
	3	♦ Multiply 2-place decimals by ones. ♦ Estimate products.	53-54	60-61	49
	4	♦ Solve word problems.	55-56	62-64	50
		Extra Practice, Unit 7, Exercise 2, pp. 121-124			
8	5	♦ Practice.	57		51
		Tests, Unit 7, 2A and 2B, pp. 27-32			
		Answers to Workbook Exercises 11-14			52
		Chapter 3: Division			53
	1	♦ Divide decimals using division facts.	58-60	65-66	54
	2	♦ Divide hundredths.	61	67	55
	3	♦ Divide decimals.	62-63	68-69	56
	4	♦ Add place values in order to divide.	63-64	70-72	57
9	5	♦ Round the quotient to 1 decimal place.	64	73	58-59
	6	♦ Solve word problems.	65-66	74-76	60
	7	♦ Practice.	67		61
		Extra Practice, Unit 7, Exercise 3, pp. 125-130			
	8	♦ Practice.	68		62
	9	♦ Practice.	69		63

Week		Objectives	Text book	Work book	Guide
Unit 11: Measures and Volume					
		Chapter 1: Adding and Subtracting Measures			123
	1	◆ Review conversion units for measurement. ◆ Convert between measurements within a measurement system.	129	144-145	124-126
	2	◆ Review addition and subtraction of measures in compound units.	128, 130	145	127-128
		Extra Practice, Unit 11, Exercise 1, pp. 169-170			
		Tests, Unit 11, 1A and 1B, pp. 171-174			
		Chapter 2: Multiplying Measures			129
17	1	◆ Multiply measures in compound units by a 1-digit number.	131-132	146-147	130
		Extra Practice, Unit 11, Exercise 2, pp. 171-179			
		Tests, Unit 11, 2A and 2B, pp. 175-180			
		Chapter 3: Dividing Measures			131
	1	◆ Divide measures in compound units by a 1-digit number.	133-134	148-149	132-133
		Extra Practice, Unit 11, Exercise 3, pp. 175-178			
	2	◆ Practice.	135-136		134
		Tests, Unit 11, 3A and 3B, pp. 181-186			
		Answers to Workbook Exercises 1-3			135
		Chapter 4: Cubic Units			136
	1	◆ Find the volume of solids made up of unit cubes. ◆ Find the volume of solids made up of 1-centimeter cubes. ◆ Find the volume of 2-dimensional representations of solids made up of unit cubes.	137-139	150	137-138
		Extra Practice, Unit 11, Exercise 4, pp. 179-180			
		Tests, Unit 11, 4A and 4B, pp. 187-197			
		Chapter 5: Volume of Rectangular Prisms			139
18	1	◆ Find the volume of rectangular prisms.	140-143	151-152	140-141
	2	◆ Practice.	145		142
	3	◆ Convert between liters and milliliters to cm^3. ◆ Solve problems involving the volume of liquids in rectangular containers.	144, 146	153-154	143-144

Week		Objectives	Text book	Work book	Guide
		Extra Practice, Unit 11, Exercise 5, pp. 181-182			
		Tests, Unit 11, 5A and 5B, pp. 199-208			
	Review 11		147-152	155-163	145-146
		Tests, Units 1-11, Cumulative A and B, pp. 209-220			
	Answers to Workbook Exercises 4-6				147
	Answers to Workbook Review 11				147-148
Answers to Mental Math					149-150
Appendix - Mental Math					a1-a7
Appendix					a8-a32

Materials

Whiteboard and Dry-Erase Markers
A whiteboard that can be held is useful in doing lessons while sitting at the table (or on the couch). Students can work problems given during the lessons on their own personal boards.

Base-10 set
A set usually has 100 unit-cubes, 10 or more ten-rods, 10 hundred-flats, and 1 thousand-block. These are used rarely at this level, so are optional.

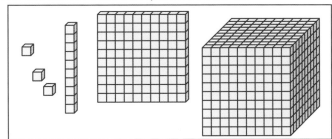

Round counters
To use as place-value discs, unless you buy commercial place-value discs, or make paper ones, or just draw them.

Place-value discs
Round discs with 0.001, 0.01, 0.1, 1, 10, or 100, written on them. You can label round counters using a permanent marker. You need 30 of each kind.

Small mirror (optional)
A mirror with a flat edge that can be placed right on the paper.

Graph paper
There are some in the appendix you can copy.

Dice or number cubes

Multilink cubes
These are cubes that can be linked together on all 6 sides.

Measurement tools
Ruler
Meter stick
Liter measuring cup
Dropper or teaspoon
Protractor
Set square

Supplements

The textbook and workbook provide the essence of the math curriculum. Some students profit by additional practice or more review. Other students profit by more challenging problems. There are several supplementary workbooks available at www.singaporemath.com.

Unit 6 – Decimals

Chapter 1 – Tenths

Objectives

♦ Understand the tenths place in decimal numbers.
♦ Express a fraction with a denominator of 10 as a decimal number.
♦ Read and write decimal numbers to one place.
♦ Locate 1-place decimal numbers on a number line.
♦ Express a 1-place decimal number as a fraction or mixed number in simplest form.
♦ Compare and order 1-place decimal numbers.
♦ Rename 10 tenths as 1 one and 1 one as 10 tenths.

Material

♦ Base-10 blocks, optional
♦ Fraction squares showing whole and tenths (appendix p. 8)
♦ Number lines (appendix p. 9)
♦ Place-value discs for 0.1, 1, and 10

Vocabulary

♦ Decimal
♦ Decimal point
♦ Tenths place

Notes

In *Primary Mathematics* 2B, students were introduced to 2-place decimal numbers in the context of money. Money is written as the number of dollars followed by a decimal point and then the number of cents. So each dollar is a whole and each cent is one hundredth. In *Primary Mathematics* 3B, students represented coins as both fractions of a dollar and with a decimal representation, giving them an introduction to some equivalent decimals and fractions. At both these levels, the decimal point in written money was called a dot. In this chapter, your student will be formally introduced to tenths, and learn how to convert between decimals and fractions.

Place-value notation makes numbers understandable, and computation accurate and simple. We use ten digits to write numbers and each digit has a value that is ten times as much as if it were in the place to the right of it and one tenth as much as if it were in the place to the left of it. The number 23,456 represents 2 ten thousands, 3 thousands, 4 hundreds, 5 tens, and 6 ones. The place value of the digit 3 is thousands, and its value is 3000. Each whole number can be expanded as the sum of multiples of the value for each place — 1, 10, 100, 1000, etc. So, 23,456 can be written as 20,000 + 3000 + 400 + 50 + 6.

Decimal numbers are an extension of place-value notation, to include place values less than 1. We write a **decimal point** to the right of the ones place. A digit in the first place to the right of the decimal point, the **tenths** place, has a value that is one tenth of the value of the same digit in the ones place. A digit in the second place to the right of the decimal point, the hundredths place, has a value of one tenth of the same digit in the tenths place. A digit in the third place to the right of the decimal point, the thousandths place, has a value of one tenth of the same digit in the hundredths place.

1.234 is $1 + 0.2 + 0.03 + 0.004$, or $1 + \frac{2}{10} + \frac{3}{100} + \frac{4}{1000}$. We can read this number as one and two hundred thirty-four thousandths, or, more simply, one point two three four.

Usually, decimal places are not called by name after the thousandths place; we do not normally say "the hundred-thousandths place" but rather the "fifth decimal place." At this level, your student will only encounter decimal numbers to the third place, or thousandths.

The digits to the left of the decimal point are whole numbers; the digits to the right of the decimal point are decimals. If a decimal number is less than 1, we usually use a 0 as a place holder in the ones place. Writing 0.12 rather than .12 makes it easier to see, and pay attention to, the decimal point.

Use concrete manipulatives to explain place-value concepts for decimal numbers, including base-ten blocks, fraction squares, number lines, and place-value discs. The example below shows 52.031 with place-value discs on a place-value chart. Place-value discs will be particularly useful in the next unit where they can be used on place-value charts to illustrate addition, subtraction, multiplication, and division. Your student should be familiar with place-value discs and charts from earlier levels of *Primary Mathematics*, and should be able to extend their use to decimals.

Tens	Ones	Tenths	Hundredths	Thousandths
10 10 10 10 10	1 1		0.0 0.0 0.01	0.001
5	2	0	3	1

(1) Read and write tenths less than 1 as a decimal number

Activity

To illustrate tenths concretely you can use base-10 blocks or similar material.

Draw a place-value chart with three columns with room to add a fourth on the right. Label the columns hundreds, tens, and ones.

Show your student the thousand-cube and ask, "What is one-tenth of 1000?" Write 100 on the place-value chart.

$\frac{1}{10}$ of 1000 = 100

Show your student the 100-flat and ask, "What is one tenth of 100?" Write 10 on the place-value chart.

$\frac{1}{10}$ of 100 = 10

Show your student the 10-rod and ask, "What is one tenth of 10?" Write 1 on the place-value chart.

$\frac{1}{10}$ of 10 = 1

Show your student the unit cube and ask: "What is one tenth of 1?" It is the fraction $\frac{1}{10}$, but because it is one tenth

$\frac{1}{10}$ of 1 = 0.1

Hundreds	Tens	Ones
1	0	0
	1	0
		1

Hundreds	Tens	Ones	Tenths
		0	1

of 1 in the ones place, we can show it in its own place value, again to the right of the ones. Add a column to the right, label it "Tenths," and write a 1 in it. Tell your student that this is the *tenths* place and to show that the digits in this place value are less than one whole, we use a dot, called a *decimal point*, to separate it from the whole numbers. We usually also put a digit in the ones place; for one tenth of 1 we put a 0 in the ones place to show that there are no ones. Tell your student that a number with a decimal point is called a *decimal*. We read the number 0.1 as either "one tenth" or as "zero point one."

Write the number 1111.1. Tell your student that each digit has a value that is one tenth of the digit to the left and ten times the digit to the right. Exactly how many items or objects the digit stands for is determined by its place. The digit after the decimal point is in the tenths place, and stands for the number of tenths of a whole.

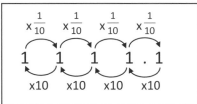

Use fraction squares showing tenths (such the ones in the appendix). Tell your student the square stands for one whole. Color in some rows and have him give the amount colored as both a fraction and a decimal number. For example, color in 3 rows. He writes $\frac{3}{10}$ and 0.3.

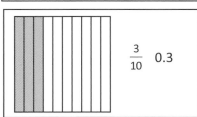

$\frac{3}{10}$ 0.3

Then write another 1-place decimal number less than 1 and have him color in the correct number of tenths.

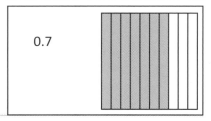

0.7

Discussion

Concept pp. 8-9

Page 8 relates tenths to measurement. Discuss the pictures and ask your student what the whole is in each picture (1 cm, 1 kg, 1 ℓ, 1 whole). Each picture shows $\frac{8}{10}$ of a whole, which can be written as 0.8. Point out that we read 0.8 as "zero point eight" or as "eight tenths."

Ask your student for the decimal for how much water we would have if we added another tenth to the beaker on p. 8. It would be 0.9 ℓ. Then ask how we would write the decimal if we added another tenth. If we add a tenth to a beaker that already has 0.9 ℓ, we now have ten tenths of a liter. Since each place can only have the digits 0 through 9, we have to go to the next larger place value, which is one. 1 one is the same as 10 tenths.

> 0.8
> + 1 tenth → 0.9
> + 1 tenth → 1.0, or 1

The top of p. 9 is an introduction to illustrating tenths with a number line. After discussing the information on the page, draw a number line. Show the 0 to 1 interval expanded, and divide it into tenths. Get your student to label each tick mark with decimals and the corresponding fraction.

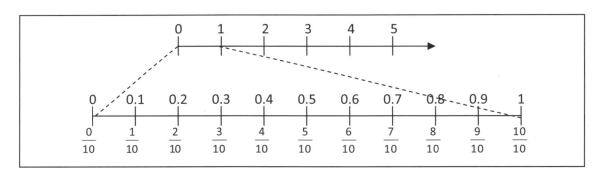

Task 1, p. 9

This task illustrates decimals with place-value discs. Have your student write and say the answers out loud.

> 1. (a) 0.4
>   (b) 0.6
>   (c) 0.9

Practice

Tasks 2-4, p. 10

Workbook

Exercise 1, pp. 7-9 (answers p. 9)

> 2. (a) 0.1   (b) 0.3
>   (c) 0.5   (d) 0.7
>
> 3. $\frac{2}{10}$
>
> 4. $\frac{4}{10}$

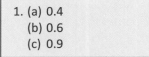

(2) Read and write tenths greater than 1 as a decimal number

Activity

Use some fraction squares showing wholes and tenths and color in some tenths, or use 100-flats and 10-rods from a base-10 set. (Tell your student that the flat is now one whole and so each rod is a tenth.) Give her some wholes and some tenths. Ask her to write the amount first as a fraction, and then as the corresponding decimal number. See the example at the right. Tell her how to read the number. The example at the right is read as "two

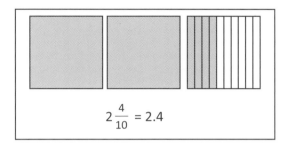

$$2\frac{4}{10} = 2.4$$

point four" or "two and four tenths." Point out that we use the word "and" between the word for ones and for tenths. 2.4 is four tenths more than 2.

$$7\frac{8}{10} = 7.8$$

Write some other mixed fractions greater than 1 with a denominator of 10, and ask your student to write them as a decimal number. Then write some improper fractions and ask him to write them as a decimal number.

$$\frac{35}{10} = 3.5$$

Use the number lines in the appendix on page a9 showing tick marks for tenths or fifths. Write some 1-place decimal numbers (such as 3.2, 6.4, and 22.9) and ask your student to locate them on the number lines. Point to a mark and have her supply the decimal number for that point.

Use a ruler. Show your student the divisions on the centimeter side. Each division is a tenth. Have him measure or draw some lengths correct to the nearest tenth and write the length as a 1-place decimal number.

On some rulers the inches from 6 inches to 12 inches are divided into tenths. Write a 1-place decimal number between 6 and 12 and have your student draw a line of the given length in inches.

If you have some beakers or graduated cylinders, you can have your student measure or read some volumes to the nearest tenth.

Practice

Tasks 5-6, p. 11

Workbook

Exercise 2, pp. 10-11 (answers p. 9)

5. (a) 0.6 cm
   (b) 0.6

6. (a) 2.5 ℓ
   (b) 1.2 kg

(3) Write a decimal number in tenths as a fraction

Activity

Use fraction squares showing wholes and tenths, or flats and rods from a base-10 set. Write a decimal number in tenths, such as 2.8, and ask your student to illustrate the number with the fraction squares or base-10 blocks. Then ask her to write the number as a mixed fraction in tenths, and then to simplify it. Repeat with a few other examples, with and without the fraction squares. Ask her to simplify the fractions whenever possible.

$$2.4 = 2\frac{4}{10} = 2\frac{2}{5}$$

$$3.3 = 3\frac{3}{10} \qquad 10.5 = 10\frac{5}{10} = 10\frac{1}{2}$$

Discussion

Tasks 7-9, p. 12

7. (a) 1.5
    (b) 2.9

8. A: 0.4   B: 0.9   C: 1.1   D: 1.6

9. (a) $\frac{1}{5}$ (b) $1\frac{1}{5}$

    (c) $\frac{4}{5}$ (d) $2\frac{4}{5}$

Activity

List all the tenths from 0.1 and have your student list the equivalent fractions in simplest form. Tell him that these are good equivalencies to memorize.

$$0.1 = \frac{1}{10} \qquad\qquad 0.2 = \frac{1}{5}$$

$$0.3 = \frac{3}{10} \qquad\qquad 0.4 = \frac{2}{5}$$

$$0.5 = \frac{1}{2} \qquad\qquad 0.6 = \frac{3}{5}$$

$$0.7 = \frac{7}{10} \qquad\qquad 0.8 = \frac{4}{5}$$

$$0.9 = \frac{9}{10} \qquad\qquad 1.0 = 1$$

Ask your student to convert some fractions where the denominator is 2 or 5 into a decimal number. Include some improper fractions. She needs to first convert to a mixed number, and then convert the fractional part by using any equivalencies she has learned, or convert the fractional part to an equivalent fraction with a denominator of 10.

$$\frac{4}{5} = \frac{8}{10} = 0.8$$

$$\frac{3}{2} = 1\frac{1}{2} = 1\frac{5}{10} = 1.5$$

$$5\frac{2}{5} = 5\frac{4}{10} = 5.4$$

$$\frac{47}{5} = 9\frac{2}{5} = 9\frac{4}{10} = 9.4$$

Refer back to the number line in Task 8, or draw a number line. Ask your student which is greater, 1.6 (D) or 0.9 (B) and why. On a number line, the numbers increase from left to right. 1.6 is to the right of 0.9 and is greater than 0.9. Show the numbers with fraction squares. Write the two numbers vertically, aligning the digits. Point out that we can compare the numbers easily by looking at each digit, starting with the one in the largest place, which in this case is ones. Since 1 one is greater than 0 ones, 1.6 is greater than 0.9.

Repeat with 1.1 (C) and 1.6 (D). 1.6 is again larger. Illustrate with fraction squares. We start by comparing the digit in the highest place (ones). They are the same, so we then compare the digit in the next lower place, which is tenths. Since 6 tenths is greater than one tenth, 1.6 is greater than 1.1.

Write some decimals for your student to put in order. You can mix in some fractions with denominators of 10.

Write a number in tenths, and ask your student to write a pattern that increases or decreases by a given number of tenths.

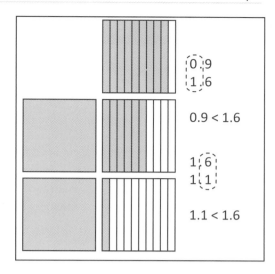

0.9
1.6

0.9 < 1.6

1.6
1.1

1.1 < 1.6

4.2   2.4 $4\frac{2}{5}$ 4

$2.4 < 4 < 4.2 < 4\frac{2}{5}$

3.5; increase by 0.3
3.5, 3.8, 4.1, 4.4, 4.7, 5, 5.3, ...

20.1; decrease by 0.5
20.1, 19.6, 19.1, 18.6, 18.1, ...

Practice

Tasks 10-13, pp. 12-13

Workbook

Exercise 3, pp. 12-13 (answers p. 9)

---

10. 3.7

11. 8.5

12. (a) 0.3, 1.3, 3, 3.1
    (b) 2.7, 7.2, 7.8, 9

13. (a) 6.8, 7.0
    (b) 9.2, 9.6

(4) Rename tenths

Activity

Use two fraction squares showing tenths. Tape them together and color 14 tenths. Or use 14 rods from a base-10 set. Ask your student to write a decimal number for it. He should write 1.4. He may write 0.14, which is incorrect. All tenths need to be in a single place value. Show him that he can trade in 10 tenths for 1 whole. If there are 10 tenths, they become a 1. 14 tenths must be renamed as 1 one and 4 tenths.

Ask:

⇒ How many tenths are there in 0.4? (4)
⇒ How many tenths are there in 1? (10)
⇒ How many tenths are there in 1.4? (14)
⇒ How many tenths are there in 2? (20)
⇒ What is the decimal number for 42 tenths? (4.2)

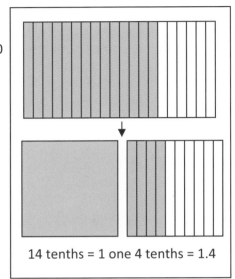

14 tenths = 1 one 4 tenths = 1.4

Use place-value discs labeled with 10, 1, and 0.1. Mix them up in a bowl or bag. Draw a place-value chart with 3 columns labeled *tens*, *ones*, and *tenths*. Have your student take a handful of discs and arrange them on the place-value chart, and write the value of the discs. If she has more than 10 of any one kind of place-value disc, she must trade it in for the next higher place value. Then rewrite the decimal number in expanded form, for example: 42.8 = 40 + 2 + 0.8.

Point out that the number is made up of two parts: the whole part, 42, and the fractional part, 0.8. The decimal point separates the whole part from the fractional part.

Repeat with other handfuls of discs as needed.

Practice

Tasks 14-15, p. 13

Workbook

Exercise 4, pp. 14-15 (answers p. 9)

Reinforcement

Write some equations such as those at the right and have your student supply the decimal number. If needed, illustrate with place-value discs.

Extra Practice, Unit 6, Exercise 1, pp. 99-100

Test

Tests, Unit 6, 1A and 1B, pp. 1-4

14. (a) 2.3	
(b) 36.5	
(c) 50.4	
15. (a) 1.2	
(b) 2.1	

10 + 0.2 = ?	10.2
0.4 + 4 + 10 = ?	14.4
6 + 0.2 + 30 = ?	36.2
$100 + \dfrac{3}{10} + 2 = ?$	102.3
30 + ? + 4 = 34.2	0.2
200 + ? + 0.4 = 243.4	43

Workbook

Exercise 1, pp. 7-9

1. (a) 0.2
   (b) 0.5
   (c) 0.8
   (d) 0.9

2. (a) 0.4 ℓ   (b) 0.7 ℓ

3. (a) 0.9 kg   (b) 0.5 kg

4. (a) 0.2   (b) 0.6
   (c) 0.9

5. (a) 0.8   (b) 0.4
   (c) 0.5   (d) 0.1

Exercise 2, pp. 10-11

1. 6.3 cm

2. (a) 6.4 cm
   (b) 9.7 cm
   (c) 8.2 cm

3. (a) 1.6 ℓ
   (b) 2.4 ℓ

4. (a) 2.8 kg   (b) 1.4 kg

Exercise 3, pp. 12-13

1.

0.1	0.2	**0.3**	**0.4**	**0.5**	0.6
$\frac{1}{10}$	$\frac{2}{10}$	$\frac{3}{10}$	$\frac{4}{10}$	$\frac{5}{10}$	$\frac{6}{10}$

1.1	1.2	**1.3**	**1.4**	2.2	**3.5**
$1\frac{1}{10}$	$1\frac{2}{10}$	$1\frac{3}{10}$	$1\frac{4}{10}$	$2\frac{2}{10}$	$3\frac{5}{10}$

2. (a) 0.4   (b) 1.4   (c) 0.5   (d) 3.5

3. (a) $\frac{3}{10}$ (b) $2\frac{3}{10}$ (c) $\frac{3}{5}$ (d) $3\frac{3}{5}$

4. (a) 0.4; 1.3; 2.8
   (b) 8.8; 10.2; 11.7
   (c) 59.5; 61.6; 64.4

5. (a) =   (b) >
   (c) =   (d) >

6. (a) 0.1
   (b) 0.9

7. (a) 6.2
   (b) 2.9

8. 5.7, 6.5, 7.3, 9.6

9. 9, 4.9, 3.6, 3.4

10. 2.7, 2.9
    6, 6.5

Exercise 4, pp. 14-15

1. (a) 34.6
   (b) 50.7
   (c) 45.3
   (d) 40.9

2. (a) 0.8
   (b) 0.3
   (c) 90
   (d) 30
   (e) 5
   (f) 9

3.

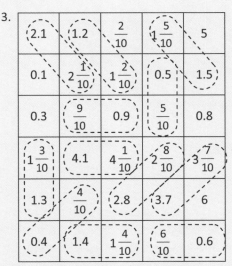

Chapter 2 – Hundredths

Objectives

- Read and write decimal numbers for hundredths.
- Relate each digit in a 2-place decimal number to its place value.
- Express a fraction or a mixed number with a denominator of a factor of 100 as a decimal number.
- Interpret and complete linear progressions involving 2-place decimals.
- Locate 2-place decimals on a number line.
- Express a decimal number as a fraction in simplest form.
- Compare and order decimal numbers to two places.
- Use mental math to add or subtract tenths or hundredths to numbers up to 2 decimal places.
- Use mental math to make a whole with hundredths.

Material

- Base-10 blocks (optional)
- Fraction squares showing whole and hundredths (appendix p. a8)
- Place-value discs for 0.01, 0.1, 1, 10, and 100
- Number lines (appendix p. a10)
- Mental Math 1-4 (appendix pp. a1-a2)

Vocabulary

- Hundredths place

Notes

In this chapter, decimal notation is extended to 2 decimal places. The first decimal place is the tenths place, and the second is the **hundredths place**. The value of the digit in the hundredths place is one tenth that of what that digit would be in the tenths place, and one hundredth that of what it would be in the ones place.

$$0.04 = \frac{4}{100} \text{ of 1 whole}$$

$$0.73 = \frac{7}{10} + \frac{3}{100} = \frac{70}{100} + \frac{3}{100} = \frac{73}{100} \text{ of 1 whole}$$

Your student will be converting decimals to fractions in simplest form. Equivalent fractions and fractions in simplest form were covered in *Primary Mathematics* 3A and 4A and factors were covered in *Primary Mathematics* 4A. At this level, your student will only be converting fractions which, in their simplest form, have denominators that are factors of 100 to 1-place or 2-place decimal numbers by simply finding an equivalent fraction with a denominator of 10 or 100. For example:

$$\frac{3}{25} = \frac{12}{100} = 0.12$$

Students will learn how to use division to convert all fractions to decimals in *Primary Mathematics* 5A.

In some other school curricula, students are taught to only read decimal numbers as fractions, using the word "and" for the decimal point. 6.58 is read as "six and fifty-eight hundredths" and

2.248 is read as "two and two hundred forty-eight thousandths." You should teach your student to recognize and use this nomenclature through thousandths, as he may encounter it on standardized tests. However, this nomenclature is cumbersome for decimals with place values larger than hundredths, and is not normally used in algebra and more advanced mathematics. In the *Primary Mathematics* curriculum, students are allowed to use the nomenclature where the decimal point is read as "point" and then the digits simply read. 2.248 is "two point two four eight." If your student already understands place value, this nomenclature should not cause any difficulties with understanding what each digit in each place stands for.

Decimal numbers are ordered in the same way as whole numbers. We start by comparing the digits in the highest place value. If they are the same, we then compare the digits in the next higher place value, and so on. As with whole numbers, we have to be careful to pay attention to the place value of the digits, not the number of digits. 4.5 is larger than 4.25 even though it has fewer digits.

In this chapter, your student will be adding or subtracting a number with a small non-zero digit to a decimal number. For example, 40.92 + 0.2 or 5.61 − 0.03. These can be done by counting up or down and the purpose is to pay attention to the place value of the digit being added or subtracted. With larger non-zero digits your student should be able to extend mental math strategies learned earlier for adding a single non-zero digit to tenths and hundredths. For example, if 68 ones + 5 ones = 73 ones, then 68 hundredths + 5 hundredths = 73 hundredths (as in 5.68 + 0.05 = 5.73) and 68 tenths + 5 tenths = 73 tenths = 7.3 (as in 6.8 + 0.5 = 7.3 or 6.81 + 0.5 = 7.31) If your student cannot do the problems mentally, you may want to review mental math strategies from *Primary Mathematics* 3A and 4A; adding a single non-zero digit mentally will not be re-taught here. She will formally learn to add and subtract decimal numbers using the standard algorithm in the next unit.

In *Primary Mathematics* 2B, students learned how to "make a 100." This was extended to making change for a dollar. For example, since 47 must be added to 53 to make 100, 0.47 must be added to 0.53 to make 1. If your student cannot make 100 mentally, you may want to review the strategies from *Primary Mathematics* 2B or 3A; they will not be re-taught here.

(1) Read and write hundredths as a decimal number

Activity

To illustrate hundredths concretely, you can use fraction squares showing a whole, tenths, and hundredths, or base-10 blocks where the 100-flat is now considered to be 1 whole.

Display a fraction square divided into ten columns and ask students what each column represents. Now show one where each column is divided into ten equal parts. Color in one square. Ask your student to compare it to the fraction square showing tenths and tell you what fraction of a tenth and of the whole it is. The colored square is 1 tenth of 1 tenth, and also 1 hundredth of 1 whole.

Write 1 on a place value chart with a column for ones, add a column for tenths and write 0.1, and then ask your student what we need to do to show a tenth of a tenth. Add a column for hundredths and write 0.01. Tell her that this new place is called the *hundredths place*.

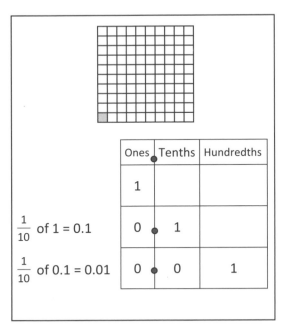

$\frac{1}{10}$ of 1 = 0.1

$\frac{1}{10}$ of 0.1 = 0.01

Ones	Tenths	Hundredths
1		
0	1	
0	0	1

Discussion

Concept pp. 14-15

The picture at the top shows a bar representing 1 whole divided into tenths, and then one of tenths further divided into 10 parts and then expanded. Ask your student how many such small parts would be in the entire bar for 1. There would be 100 small parts. Each small part is a tenth of a tenth. Three tenths and seven hundredths are shaded, or 0.37.

Point out the two methods for reading a decimal. Be sure your student understands why 0.37 = 37 hundredths = 3 tenths 7 hundredths. 0.37 is 30 hundredths and 7 hundredths. From the image, he can see that each ten hundredths is the same as a tenth.

Activity

Place-value discs are a more abstract representation of base-10 concepts than fraction squares or number lines. Give your student a 1-disc and ask her to trade it in for 0.1-discs. She should trade it in for ten 0.1-discs. Then ask her to trade in one 0.1-disc for 0.01-discs. Again, a 0.1-disc has the same value as ten 0.01-discs. She should now have nine 0.1-discs and ten 0.01-discs. Ask her to write the number these discs represent. They represent the number 1. If needed, have her place the discs on a place-value chart. Since we cannot write a single digit to represent the ten 0.01 discs, we have to rename them as 0.1, and then the ten 0.1-discs will need to be renamed as 1.

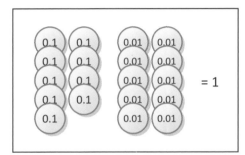

= 1

Give your student a handful of 10-discs, 1-discs, 0.1-discs, and 0.01-discs and have him determine and write down the number they represent, trading in ten 0.01-discs for a 0.1-disc, ten 0.1-discs for a 1-disc, and ten 1-discs for a 10-disc as needed. After he has written the number, ask him what place each digit is in, and what the value of each digit is. For example, if the resulting number is 10.93, the value of the digit 9 is 9 tenths or 0.9.

> 10.93
>
> 1: tens place, value 10
> 0: ones place, value 0
> 9: tenths place, value 0.9
> 3: hundredths place, value 0.03

Practice

Tasks 1-4, pp. 15-16

1-2: In these tasks, the number discs are pictorial. Ask your student to also write the fraction or mixed number with 100 in the denominator for each problem.

For Task 1(c), point out that the ten hundredths are renamed as a tenth. $\frac{10}{100}$ is the same as $\frac{1}{10}$, so $\frac{12}{100} = \frac{1}{10} + \frac{2}{100}$. For Task 2(a) point out that the 0 in the tenths place means that there are no tenths.

3: In this task, the discs are shown on a place-value chart.

4: In this task, the discs on the chart are replaced with just numbers.

Write down a number without a place-value chart and ask your student for the value of each digit. For example:

205.04:

What digit is in the tenths place?	(0)
What is the value of the digit 2?	(200)
What is the value of the digit 4?	(4 hundredths)

> 1. (a) 0.03, $\frac{3}{100}$ (b) 0.05, $\frac{5}{100}$
>
>   (c) 0.12, $\frac{12}{100}$
>
> 2. (a) 3.02, $3\frac{2}{100}$
>
>   (b) 4.25, $4\frac{25}{100}$
>
> 3. 3: 30
>   4: 4
>
>   5: 0.5 or $\frac{5}{10}$
>
>   6: 0.06 or $\frac{6}{100}$
>
> 4. 9: 0.9 or $\frac{9}{10}$
>
>   2: 0.02 or $\frac{2}{100}$
>   7: 7
>   4: 40
>   3: 300

Workbook

Exercise 5, pp. 16-18 (answers p. 19)

(2) Convert hundredths to decimal numbers

Discussion

Tasks 5-6(b), pp. 17-18

5(a): This task shows that 40 hundredths is the same as 4 tenths. Each little square is a hundredth, each column of 10 little squares is a tenth, so 40 little squares is the same as 4 tenths. Point out that we can write $\frac{40}{100}$ as 0.40, but this is the same as 0.4.

5(b): Point out that in the second square, two columns are colored. You can break out the fractional part as the sum of tenths and hundredths. $\frac{28}{100} = \frac{20}{100} + \frac{8}{100} = \frac{2}{10} + \frac{8}{100} = 0.2 + 0.08$.

5(c): Make sure your student understands that other than the two wholes, there are no tenths, so the decimal number has a 0 in the tenths place.

6(a): Remind your student that we write money as the number of dollars, a dot, and the number of cents, allowing two places for cents. The dot is the same as a decimal point; the dollar is the whole. There are 100 cents in a dollar, so each cent is $\frac{1}{100}$ of a dollar, or $0.01.

6(b): 10 cents is the same as $\frac{10}{100}$ of a dollar, or $0.10. Point out that 10 cents is also a tenth of a dollar, but with money the two decimal places represent the number of cents, and so we always use both decimal places after the dot; we do not write 10 cents as $0.1.

Practice

Tasks 6(c)-7, p. 18

Workbook

Exercise 6, pp. 19-20 (answers p. 19)

---

5. (a) 0.4
   (b) 1.28
   (c) 2.05

6. (a) $\frac{1}{100}$ (b) $\frac{1}{10}$

---

6. (c) $\frac{2}{10}$ (d) $\frac{5}{10}$
     20¢ $0.50

   (e) $\frac{45}{100}$ (f) $\frac{26}{100}$
     45¢ $0.26

7. (a) $3.85   (b) $6.50
   (c) $8.05   (d) $85.00

(3) Solve problems involving decimals and fractions

Activity

Use place-value discs labeled with 1, 0.1, and 0.01, about 30 of each, if possible. Mix them up in a bowl or bag. Have your student pick up a handful of discs, write how many she has of each kind, determine the value of the discs, and write the decimal number. If necessary, she can trade in ten of one kind for one of the next higher place and use a place-value chart. However, she may be able to do this activity without physically trading in discs. Repeat with other handfuls of discs as needed.

> 3 ones + 15 tenths + 21 hundredths
> = 3 + 1.5 + 0.21
> = 4.71
> Or (renaming first)
> 3 ones + 15 tenths + 21 hundredths
> = 4 + 0.7 + 0.01
> = 4.71

Practice

Tasks 8-10, pp. 18-19

9: If your student has trouble with this task, you can also illustrate what is happening with place-value discs. For example, for 9(a) set out 3 ones and add three tenths 5 times, renaming by trading in as needed, and asking him for the value each time.

> 8. (a) 2.84   (b) 36.25
>   (c) 54.03   (c) 80.57
>
> 9. (a) 3.9, 4.2
>   (b) 8.4, 8.2
>   (c) 4.75; 4.85
>
> 10. (a) A: 0.04   B: 0.07   C: 0.11
>   D: 0.13   E: 0.19
>   (b) P: 4.62   Q: 4.66   R: 4.69
>   S: 4.73   T: 4.78

Workbook

Exercise 7, pp. 21-22 (answers p. 19)

Reinforcement

Write some equations such as the following and have your student supply the decimal number.

♦ 0.44 + 4 + 40 = ?   44.44

♦ $6 + 0.2 + 30 + \dfrac{1}{100} = ?$ 36.21

♦ $100 + \dfrac{3}{10} + 0.02 + 8 = ?$ 108.32

♦ $\dfrac{34}{100} + 100 = ?$ 100.34

♦ 450 tenths = ?   45

♦ 450 hundredths = ?   4.5

♦ 3.33 = ? hundredths   333

Discuss the following patterns with your student and have her supply the next two numbers.

♦ 0.4, 0.6, 0.8, _____, _____   1, 1.2

♦ 3.5, 3, 2.5, 2, 1.5, _____, _____   1, 0.5

♦ 0.07, 0.1, 0.13, _____, _____   0.16, 0.19

♦ 0.25, 0.5, 0.75, _____, _____   1, 1.25

Use the number lines on appendix p. a10. Ask your student to mark the position of some decimal numbers you tell him. For example, ask him to mark and label the position for 1.15 on the second number line (between two tick marks).

(4) Convert between 2-place decimals and fractions

Discussion

Task 11, p. 19

11(a): After your student has simplified the fraction, ask her what coin could be $0.25. (A quarter). Since there are four quarters in a dollar, then one quarter is a fourth of a dollar. Ask her to use this idea to find 0.75 as a fraction: $0.75 = 3 quarters = $\frac{3}{4}$ of a dollar.

> 11. (a) $\frac{1}{4}$
>
> (b) $1\frac{21}{25}$

Practice

Task 12, p. 20

After your student has finished this task, ask him to find the factors of 100 and compare them to the denominators in the answers. The factors of 100 are 1, 2, 4, 5, 10, 20, 25, 50, and 100. All decimal numbers to hundredths rewritten as fractions in their simplest form will have only those numbers as denominators. Ask him if he can think of some decimal numbers that will have 100 as the denominator of the fraction in simplest form (those that do not have a common factor with 100, e.g. 0.49). Ask him what decimal numbers might have 20 as the denominator of the simplified fractions (those with a 5 in the hundredths place, e.g. 0.45).

> 12. (a) $\frac{3}{50}$ (b) $\frac{7}{25}$ (c) $\frac{6}{25}$
>
> (d) $2\frac{1}{20}$ (e) $3\frac{13}{20}$ (f) $4\frac{3}{4}$

Discussion

Task 13, p. 20

This task illustrates that we have to first convert the fraction to an equivalent fraction with 10 or 100 in the denominator in order to express it as a decimal.

> 13. (a) $\frac{6}{10}$ = **0.6**
>
> (b) $\frac{45}{100}$ = **0.45**

Practice

Task 14, p. 20

After this task, list the fractions at the right and ask your student to express each as a decimal. Tell him that if he memorizes these, then it will be easy to convert them and their multiples to a decimal. For example, since $\frac{1}{5}$ = 0.2, then $\frac{3}{5}$ = 3 x 0.2 = 0.6. Use Mental Math 1.

> 14. (a) 0.75   (b) 0.35   (c) 0.32
> (d) 1.5   (e) 2.4   (f) 3.54

> $\frac{1}{2}$ = 0.5 $\frac{1}{50}$ = 0.02
>
> $\frac{1}{4}$ = 0.25 $\frac{1}{25}$ = 0.04
>
> $\frac{1}{5}$ = 0.2 $\frac{1}{20}$ = 0.05
>
> $\frac{1}{10}$ = 0.1 $\frac{1}{100}$ = 0.01

Workbook

Exercise 8, pp. 23-24 (answers p. 19)

(5) Compare and order decimal numbers

Discussion

Tasks 15-16, p. 21

15(a): Make sure your student realizes that even though 2.9 has fewer digits, it is larger. The ones are the same, so we compare the tenths. 9 tenths (in 2.9) is greater than 1 tenth (in 2.12).

15(b): All we need to compare is the ones. Remind your student that we start with the digits in the highest place in order to compare numbers.

16(a): Ask your student to explain why 562.41 is greater than 562.38. Hundreds, tens, and ones of both digits are the same, but 4 tenths is greater than 3 tenths, so 562.41 is greater.

16(b): Again, ask your student to explain his answer. Make sure he understands that even though both numbers have 4 digits, and 243.5 starts with a smaller digit than 89.67, it is larger because it is hundreds, and 89.67 has no hundreds.

| 15. (a) < |
| (b) < |
| 16. (a) 562.41 |
| (b) 89.67 |

Practice

Tasks 17-18, p. 22

If necessary, have your student rewrite the problems one on top of the other, or in a place-value chart, aligning the digits according to place value.

| 17. (a) 42.6 |
| (b) 2.5 m |
| (c) 32.6 kg |
| 18. (a) 2.2, 2.02, 0.2, 0.02 |
| (b) 80.7, 74.5, 7.8, 7.45 |

Workbook

Exercise 9, pp. 25-26 (answers p. 19)

Reinforcement

Have your student put the numbers at the right in ascending order.

| 10.02, 10.25, 10.2, 12.5, 12.05, 1.25 |
| Answer: |
| 1.25, 10.02, 10.2, 10.25, 12.05, 12.5 |

Game

Material: Place-value discs in a bag (100's, 10's, 1's, 0.1's, and 0.01's).

Procedure: Each player draws 10 discs without looking in the bag and writes down the number formed. The player with the greatest number gets a point. The winner is the one who gets 10 points (or some other target number) first.

(6) Add or subtract in one place

Activity

If necessary, you can use place-value discs, or have your student draw place-value discs to illustrate the problems in this activity. Have her show the first number with the discs, and then add and subtract the second by adding or removing discs from the appropriate place, renaming when needed. If your student already has a good grasp of place value, or you have done some of the reinforcement activities with place-value discs from the previous lessons, she may not need to represent the numbers concretely. She can use mental math strategies, or count up or back in the appropriate place.

Have your student find the answers to the problems shown at the right. Discuss the place value of the non-zero digit being added or subtracted, which digit will change in the first number, and whether renaming occurs. For example, in the first problem, we are adding 2 hundredths and in the second 2 tenths.

5.77 + 0.02	(5.79)
5.77 + 0.2	(5.97)
5.77 + 0.03	(5.80)
5.77 + 0.3	(6.07)
5.33 − 0.02	(5.31)
5.33 − 0.2	(5.13)
5.33 − 0.04	(5.29)
5.33 − 0.4	(4.93)

Practice

Tasks 19-24, p. 22

Workbook

Exercise 10, pp. 27-28 (answers p. 19)

Reinforcement

Extra Practice, Unit 6, Exercise 2, pp. 101-104

Mental Math 2-4

Test

Tests, Unit 6, 2A and 2B, pp. 5-8

19. (a) 412.44
    (b) 412.24

20. (a) 123.49
    (b) 123.47

21. (a) 5.17
    (b) 28.6

22. (a) 86.63   (b) 24.85   (c) 4.89
    (d) 54.22   (e) 6.2   (f) 3.43

23. (a) 0.54   (b) 0.04

24. 0.18

Workbook

Exercise 5, pp. 16-18

1. (a) 0.82 (b) 8.34 (c) 3.05
   (d) 5.17 (e) 20.09

2. (a) 34.02 (b) 40.25 (c) 24.13 (d) 30.04

3. (a) 0; 0 (b) 0; 0
   (c) tenths; 0.4 (d) tens; 50
   (e) hundredths; 0.03 (f) ones; 0

4. (a) 0.03 (b) 0.01 (c) 0.09 (d) 0.8
       0.2   0.4   0   8
       0   7   6   10
       90   80   50   200

Exercise 6, pp. 19-20

1. (a) 0.07 (b) 1.07
   (c) 0.58 (d) 2.58
   (e) 0.24 (f) 1.24
   (g) 0.65 (h) 3.65
   (i) 0.03 (j) 2.03
   (k) 0.05 (l) 10.05

2. $\frac{9}{10} \to 0.9$ $\frac{17}{100} \to 0.17$ $\frac{7}{100} \to 0.07$

   $\frac{3}{10} \to 0.3$ $\frac{29}{100} \to 0.29$ $\frac{7}{10} \to 0.7$

   $\frac{9}{100} \to 0.09$

Exercise 7, pp. 21-22

1. (a) 80.7 (b) 24.5
   (c) 34.04 (d) 7.29

2. (a) $\frac{7}{100}$ (b) $\frac{5}{100}$ (c) $\frac{2}{10}$

   (d) $\frac{7}{10}$ (e) $\frac{4}{10}$

3. (a) 0.04 (b) 0.05 (c) 0.1
   (d) 0.08 (e) 0.3

4. (a) 1; 1.2 (b) 3, 3.5
   (c) 2.7; 2.5 (d) 8.5; 7.5
   (e) 0.2; 0.3 (f) 0.3, 0.25
   (g) 0.08; 0.12 (h) 9.85; 9.75

5. (a) 0.13; 0.28
   (b) 0.87; 0.97
   (c) 3.08; 3.22; 3.37

Exercise 8, pp. 23-34

1. (a) $\frac{1}{2}$ (b) $2\frac{1}{2}$ (c) $\frac{2}{25}$ (d) $1\frac{2}{25}$

   (e) $\frac{3}{20}$ (f) $3\frac{3}{20}$ (g) $\frac{16}{25}$ (h) $1\frac{16}{25}$

2. $\frac{2}{10}$, 0.2

3. $\frac{75}{100}$, 0.75

4. (a) $\frac{5}{10} = 0.5$ (b) $3\frac{5}{10} = 3.5$

   (c) $\frac{6}{10} = 0.6$ (d) $1\frac{6}{10} = 1.6$

   (e) $\frac{25}{100} = 0.25$ (f) $2\frac{25}{100} = 2.25$

   (g) $\frac{16}{100} = 0.16$ (h) $1\frac{16}{100} = 1.16$

5. (a) 0.8 (b) 3.8
   (c) 0.45 (d) 1.45
   (e) 0.06 (f) 2.06

Exercise 9, pp. 25-26

1. (a) > (b) > (c) < (d) >

2. (a) < (b) > (c) < (d) >
   (e) = (f) > (g) = (h) >

3. (a) 0.88 (b) 2.99 (c) 0.42

4. (a) 3 (b) 8.1 (c) 7.01

5. (a) 6.1, 6.01, 1.06, 0.61
   (b) 5.33, 5.3, 5.03, 5

Exercise 10, pp. 27-28

1. (a) 324.57 (b) 234.05

2. (a) 46.15 (b) 39.21 (c) 59.98 (d) 42.49
   (e) 0.1 (f) 0.01 (g) 0.1 (h) 0.01

3. (a) 5.56 (b) 4.95 (c) 4.02 (d) 7.23
   (e) 4.58 (f) 8.1 (g) 6.5 (h) 5.34

4. (a) 2.33 (b) 4.68
   (c) 3.98 (d) 1.64
   (e) 3.45 (f) 4.22
   (g) 5.19 (h) 3.63

5. (b) 0.38 (c) 0.99 (d) 0.92

Chapter 3 – Thousandths

Objectives

- Read and write decimal numbers for thousandths.
- Interpret 3-place decimals in terms of place value.
- Count up or down by a hundredth or a thousandth.
- Compare and order decimal numbers.
- Write a decimal number as a fraction in simplest form.
- Compare and order a mixture of whole numbers, decimals, and fractions.

Material

- Place-value discs for 0.001, 0.01, 0.1, 1, 10, and 100
- Mental Math 5-6 (appendix p. a2)

Vocabulary

- Thousandths place

Notes

In this chapter, decimal notation is extended to 3 decimal places. The third decimal place is the **thousandths place**. 1 one = 1000 thousandths, 1 tenth = 100 thousandths, and 1 hundredth = 10 thousandths.

$$0.125 = \frac{125}{1000} \text{ of 1 whole}$$

$$0.125 = 0.1 + 0.02 + 0.005 = \frac{1}{10} + \frac{2}{100} + \frac{5}{1000}$$

Your student will convert 3-place decimal numbers to fractions, using the same strategies he used with 1-place and 2-place decimal numbers, and strategies he has already learned for finding equivalent fractions. The resulting fractions will have denominators which are factors of 1000: 2, 4, 5, 8, 10, 20, 25, 40, 50, 100, 125, 200, 250, 500, and 1000. He will not have to convert fractions to decimals that result in 3-place decimals (except in an enrichment exercise). All the fractions he has to convert will result in decimal numbers of no more than 2 places at this level.

Your student will also compare and order decimal numbers of up to 3 decimal places and fractions. To compare fractions and decimals, she will generally need to convert the fractions to decimals.

For your information, extra zeroes are written after a decimal number to indicate the precision of a measurement. For example, a measurement of 0.020 m means that the length was measured to the nearest millimeter; it is between 0.0195 m and 0.0204 m. 0.02 m means that the length was measured to the nearest centimeter; it is between 0.015 m and 0.024 m.

(1) Read and write thousandths as a decimal number

Activity

Ask your student to write the number (not a fraction) for one tenth of 10, and name the number, then for one tenth of that, and name the number, then for one tenth of that, and name the number. He may be able to write and name one thousandth on his own. Tell him that in order to write a number for a tenth of a hundredth, we add another place to the right. This is called the *thousandths place*, because there are 1000 thousandths in one whole. Have your student write the fractions for one tenth, one hundredth, and one thousandth.

10	ten
1	one
0.1	one tenth
0.01	one hundredth
0.001	one thousandth

Tell your student that we could keep on adding new places after the decimal, each one a tenth of the previous one. You may want to discuss with her why we need such small parts of a whole. One place they are used is in measurement. A centimeter is one hundredth of a meter, so a centimeter is the same as 0.01 meters. A millimeter is one thousandth of a meter, and is the same as 0.001 meters. Scientists measure things even smaller than a millimeter. A dust mite is 400 microns long. A micron is one thousandth of a millimeter, and can be written as 0.000001 m, so a dust mite is 0.0004 m long.

Use place-value discs. Show your student the 0.001-disc and ask him how many are in a 0.01-disc (10), a 0.1-disc (100), and a 1-disc (1000). If you have been using base-10 blocks and calling the flat a whole, then a thousandth would be one tenth of the unit cube (which is now a hundredth).

1 hundredth = 10 thousandths
1 tenth = 100 thousandths
1 one = 1000 thousandths

Give your student 15 of the 0.001-discs and ask her to write the decimal number. She can rename 10 thousandths as 1 hundredth. Then give her twelve 0.001-discs and twelve 0.01-discs and ask her to write the decimal number.

15 thousandths
= 1 hundredth + 5 thousandths
= 0.01 + 0.005 = 0.015
12 hundredths + 12 thousandths
 = 1 tenth + 2 hundredths
         + 1 hundredth + 2 thousandths
= 0.132

Write the fractions shown at the right, and write the decimal numbers as shown. Each is the number of thousandths, so twenty thousandths is 0.020, but that is the same as 0.02. The number of zeros after the last non-zero digit in a decimal does not affect the value of the decimal number.

$\frac{2}{1000} = 0.002$

$\frac{20}{1000} = 0.020 = 0.02$

$\frac{200}{1000} = 0.200 = 0.2$

$\frac{2000}{1000} = 2.000 = 2$

Discussion

Concept p. 23
Tasks 1-3, pp. 23-24

Workbook

 Exercise 11, pp. 29-30 (answers p. 26)

Reinforcement

Use place-value discs (1's, 0.1's, 0.01's, and 0.001's) in a bag and a place-value chart. Have your student draw about 10 of them, and arrange them on the chart. Then have him write a decimal number, a mixed fraction with 1000 in the denominator, the sum of the value of each digit, and the sum of fractions with 10, 100, or 1000 in the denominator. For example,

4 ones, 3 tenths, 2 hundredths, 2 thousandths

4.322

$4\dfrac{322}{1000}$

4 + 0.3 + 0.02 + 0.002

$4 + \dfrac{3}{10} + \dfrac{2}{100} + \dfrac{2}{1000}$

For more challenge, have your student draw a larger handful so that she is likely to get more than 10 of any type of disc and will have to rename some.

Then ask him to write the number that is 0.01 more, 0.01 less, 0.001 more, and 0.001 less than the given number. For a bit more challenge, have him write the number that is 0.03 more, 0.03 less, 0.003 more, and 0.003 less than the given number.

Enrichment

Write the problems at the right and have your student use mental math strategies to solve them. For example, in 5.555 + 0.005, since 55 and 5 is 60, then 0.055 and 0.005 is 0.06, so 5.5_55_ + 0.00_5_ = 5.5_6_. Use Mental Math 5-6 for more practice.

(a) 0.024
(b) 0.315
(c) 4.002

1. (a) five thousandths, 0.005
   (b) 2: 20, 2 tens
       0: 0, 0 ones
       4: 0.4, 4 tenths
       3: 0.03, 3 hundredths

2. (a) 5.63
   (b) 5.61
   (c) 4.537
   (d) 4.535

3. (a) 0.148
   (b) 0.048
   (c) 0.008

5.5_55_ + 0.00_5_	(5.5_6_)
5.5_55_ + 0.0_6_	(5._6_15)
5.5_55_ + 0._7_	(_6_.255)
5.5_55_ − 0.00_8_	(5.5_47_)
5.5_55_ − 0.0_9_	(5._465_)
5.5_55_ − 0._5_	(5._0_55)

(2) Compare and order decimal numbers

Discussion

Task 4, p. 24

As with any other set of numbers to compare, we compare 3-place decimal numbers by comparing the digits in the highest place-value first. Get your student to explain her answers. In (a) we compare the digits in the tenths place to find which is greater, and in (b) we compare the digits in the thousandths place.

> 4. (a) 42.54
> (b) 63.182

Practice

Tasks 5-6, pp. 24-25

If necessary, your student can rewrite the numbers vertically, aligning the digits.

> 5. (a) 3.02, 0.32, 0.302, 0.032
> (b) 2.628, 2.189, 2.139, 2.045
>
> 6. (a) 0.538, 0.83, 3.58, 5.8
> (b) 9.047, 9.067, 9.074, 9.076

Workbook

Exercise 12, p. 31 (answers p. 26)

Game

Material: Place-value discs in a bag (10's, 1's, 0.1's, 0.01's, and 0.001's).

Procedure: Each player draws 10 discs without looking in the bag and writes down the number formed. The player with the greatest number gets a point. The winner is the one who gets 10 points (or some other target number) first.

(3) Convert decimals to fractions

Activity

Write 1.032 as a decimal number and as a mixed number, and then discuss how the fraction part can be simplified. It can be simplified in one step by dividing the numerator and denominator by 8, or in several steps, such as those shown at the right, where the numerator and denominator are successively divided by 2.

$$1.032 = 1\frac{32}{1000} = 1\frac{16}{500} = 1\frac{8}{250} = 1\frac{4}{125}$$

$$0.125 = \frac{125}{1000} = \frac{25}{200} = \frac{5}{40} = \frac{1}{8}$$

$$0.272 = \frac{272}{1000} = \frac{136}{500} = \frac{68}{250} = \frac{34}{125}$$

Ask your student to write 0.125 and 0.272 as fractions in simplest form. For 0.125, we can divide by 5 continuously.

Tell your student that $0.125 = \frac{1}{8}$ is good to memorize, since knowing that 125 x 8 = 1000 will be useful later for mental math.

Help your student list the factors for 1000: 1, 2, 4, 5, 8, 10, 20, 25, 40, 50, 100, 125, 200, 250, 500, and 1000. Disregarding 1, all 3-place decimal numbers expressed as a fraction will have one of these in the denominator. They are all multiples of 5 or 2, so to simplify the fraction with 1000 in the denominator we only need to divide by 2 or 5; it will not be of any use to try other factors such as 3.

Practice

Tasks 7-11, p. 25

Workbook

Exercise 13, pp. 32-33 (answers p. 26)

Reinforcement

Ask your student to put the following numbers in increasing order.

$$3\frac{1}{5}, 2.309, 30.29, 2\frac{39}{100}$$

They can all be changed to decimal numbers:

$$3.2, 2.309, 30.29, 2.39$$

and then put in order. Here, though, it is not necessary to convert all of them. We can first look at the whole number parts and order them by the whole number.

Then, only 2.309 and $2\frac{39}{100}$ need to be compared. The order is

$$2.309, 2\frac{39}{100}, 3\frac{1}{5}, 30.29$$

7. $\frac{13}{250}$

8. (a) $\frac{1}{2}$ (b) $\frac{2}{25}$

   (c) $\frac{3}{125}$ (d) $\frac{69}{200}$

9. $2\frac{9}{200}$

10. (a) $2\frac{3}{5}$ (b) $6\frac{1}{20}$

    (c) $3\frac{1}{500}$ (d) $2\frac{51}{125}$

11. (a) $0.6, 0.652, \frac{4}{5}, 2$

     $(\frac{4}{5} = 0.8)$

   (b) $\frac{7}{25}, 0.35, 1\frac{3}{4}, 7.231$

     $(\frac{7}{25} = 0.28, 1\frac{3}{4} = 1.75)$

Extra Practice, Unit 6, Exercise 3, pp. 105-108

Practice

Practice A, p. 26

1. (a) 0.6   (b) 6   (c) 0.06   (d) 0.006

2. (a) 4
   (b) 7

3. (a) 5.509
   (b) 2.819
   (c) 13.52

4. (a) 0.72   (b) 3.78
   (c) 5.8   (d) 8.04

5. (a) $\frac{2}{25}$ (b) $\frac{7}{50}$ (c) $\frac{29}{200}$ (d) $\frac{51}{125}$

   (e) $3\frac{3}{5}$ (f) $1\frac{3}{25}$ (g) $4\frac{253}{500}$ (h) $2\frac{3}{500}$

6. (a) 0.9   (b) 0.03   (c) 0.039   (d) 0.105
   (e) 1.7   (f) 2.18   (g) 3.007   (h) 0.999

7. (a) 0.07   (b) 0.005
   (c) 2   (d) 1000

Practice B, p. 27

1. (a) 0.008, 0.009, 0.08, 0.09
   (b) 3.025, 3.205, 3.25, 3.502
   (c) 4.386, 4.638, 4.683, 4.9
   (d) 9.392, 9.923, 9.932, 10

2. (a) 0.5   (b) 0.75   (c) 0.2
   (d) 3.8   (e) 6.25   (f) 4.6

3. (a) =   (b) >
   (c) <   (d) =
   (e) >   (f) >

4. (a) 1.703   (b) 0.085
   (c) 5.069   (d) 10.052

5. (a) 0.248   (b) 0.792
   (c) 3.78   (d) 10.504
   (e) 7.009   (f) 9.803

Tests

Tests, Unit 6, 3A, pp. 9-10

Workbook

Exercise 11, pp. 29-30

1. (a) 0.004
   (b) 4.007
   (c) 0.083
   (d) 0.435

2. (a) 0.003
   (b) 0.406

3. (a) $\frac{9}{1000}$

   (b) $\frac{43}{1000}$

4. (a) 3 ones, 4 tenths, 7 hundredths, 9 thousandths
   (b) 4; 0.4
   (c) 0.009
   (d) 0.07

5. (a) 8.4, 8.8, 9.1, 9.5
   (b) 3.22, 3.25, 3.29, 3.32
   (c) 5.999, 6.002, 6.007, 6.012
   (d) 5.265, 5.269, 5.272, 5.275

Exercise 12, p. 31

1. (a) 4.7   (b) 9.1
   (c) 1.924   (d) 5

2. (a) 624.8
   (b) 5.73
   (c) 1.1

3. (a) >
   (b) <
   (c) =
   (d) >
   (e) >
   (f) <

4. 2.128, 2.18, 2.218, 2.8

5. 6.952, 6.3, 6.295, 6.03

Exercise 13, pp. 32-33

1. (a) $\frac{16}{25}$ (b) $\frac{19}{50}$

   (c) $2\frac{2}{25}$ (d) $4\frac{19}{20}$

   (e) $\frac{27}{125}$ (f) $\frac{44}{125}$

   (g) $3\frac{88}{125}$ (h) $2\frac{17}{40}$

2. (a) 2.75
   (b) 0.5
   (c) $1\frac{1}{2}$
   (d) 0.65

3. (a) 1.245, 1.254, 1.425, 1.524
   (b) 0.097, 0.119, 0.19, 0.91
   (c) $1\frac{9}{10}$, 2.5, $3\frac{1}{2}$, 3.95
   (d) 7.1, $7\frac{1}{5}$, 7.5, $7\frac{3}{5}$

Chapter 4 – Rounding

Objectives

♦ Round decimal numbers to the nearest whole number.
♦ Round decimal numbers to the nearest tenth.

Notes

In *Primary Mathematics* 3A, students learned to round numbers of up to 4 digits to the nearest ten, hundred, or thousand. In *Primary Mathematics* 4A, students learned to round larger numbers to the nearest hundred, thousand, ten thousand, hundred thousand, and million. In this chapter, your student will learn to round decimal numbers to the nearest whole number or tenth.

Being able to round numbers to a certain place can be used to approximate the answers to addition, subtraction, multiplication, or division problems. Quick approximations are very useful with decimal numbers since it is easy to make a mistake in placing the decimal correctly in the answer. For example, the process for multiplying 42.4 x 0.7 is the same as for 424 x 7, but the answer is 29.68 instead of 2968. We can check the placement of the decimal by rounding the factors to get an approximate answer: 40 x 1 = 40, so the answer cannot be 2.968 or 296.8. Rounding will also be used in division for decimal quotients. For example, 46 ÷ 7 = 6.6 to the nearest tenth (the actual answer is a non-terminating decimal). In later levels, students will also be using estimates of irrational numbers, such as 3.14 for π and will be rounding answers for area and circumference of a circle.

By convention, if a number is exactly halfway between the place the number is being rounded to, it is rounded to the higher number. For example, 465 rounded to the nearest ten is 470.

We can follow the same process in rounding a decimal number as was used in rounding a whole number. To round a number to a specified place, we look at the digit in the next lower place. If it is 5 or greater than 5, we round up. If it is smaller than 5, we round down.

Round 24.25 to the nearest whole number:

24.25 → 24   2 in the tenths place; the number is closer to 24 than 25.

Round 24.25 to the nearest tenth:

24.25 → 24.3   5 in the hundredths place; round up.

Students have already used number lines to understand the process of rounding. They are used here as well so that your student understands the process concretely. Even locating a number on a number line requires approximation, so you may want to include activities that require your student to locate a number on a number line.

(1) Round to the nearest whole number

Discussion

Concept p. 28
Tasks 1-3, p. 29

> As you discuss rounding these numbers using their positions on the number lines, list each of them, underline the number we are rounding to, and write the rounded number. After you are finished with Task 3, have your student look at the list. Point out that to round a number to a given place, we can look at the number to the right of that place. If it is 5 or up, we round up, if it is less than 5, we round down.

1. 37 kg
2. 6 m
3. 25

16<u>4</u>.3 → 164
3<u>7</u>.4 → 37
<u>5</u>.78 → 6
2<u>4</u>.5 → 25

Practice

Task 4, p. 30

4. (a) 4   (b) 14 (c) 30   (d) 5   (e) 16 (f) 19

Workbook

Exercise 14, pp. 34-35 (answers p. 31)

Reinforcement

Draw a number line and mark whole numbers at evenly spaced intervals. Ask your student to locate various 1-place and 2-place decimals on it. To do so, he needs to estimate how close they are to the whole numbers. For example, if the number line is marked from 1-10 and you ask him to locate 2.43, he will need to realize it is closer to 2 than 3, close to halfway between them.

Enrichment

Ask your student for the set of 1-place decimals (tenths) that can be rounded to a given whole number, such as 20.

Tenths that can be rounded to 20: 19.5, 19.6, 19.7, 19.8, 19.9, 20, 20.1, 20.2, 20.3, 20.4

Ask your student to round 0.2 m to the nearest meter.

0.2 m rounded to the nearest meter is 0 m.

(2) Round to the nearest tenth

Discussion

Tasks 5-6, p. 30

As you discuss rounding these numbers using their positions on the number lines, list each of them, underline the number we are rounding to, and write the rounded number. After you are finished with the task, discuss how to round the numbers without a number line.

> 5. (a) 3 m
> (b) 3.2 m
> 6. (a) 4.3
> (b) 4.3
> (c) 4.4

> 3.18 → 3
>
> 3.18 → 3.2
>
> 4.26 → 4.3
>
> 4.32 → 4.3
>
> 4.35 → 4.4

Ask your student to round 394.65 to the nearest hundred, ten, one, and tenth, without a number line.

> 394.65 → 400
>
> 394.65 → 390
>
> 394.65 → 395
>
> 394.65 → 394.7

Practice

Task 7, p. 30

> 7. (a) 0.9   (b) 2.5   (c) 7.1
> (d) 11.0   (e) 18.0   (f) 24.6

Workbook

Exercise 15, p. 36 (answers p. 31)

Reinforcement

Extra Practice, Unit 6, Exercise 4, pp. 109-110

Enrichment

Ask your student for the set of 2-place decimals that can be rounded to a given 1-place decimal, such as 6.5.

> Hundredths that can be rounded to 6.5:
> 6.45, 6.46, 6.47, 6.48, 6.49,
> 6.51, 6,52, 6.53, 6.54

Test

Tests, Unit 6, 4A and 4B, pp. 11-13

Review 6

Review

Review 6, pp. 31-34

Workbook

Review 6, pp. 37-41 (answers p. 31)

Tests

Tests, Units 1-6 Cumulative Tests A and B, pp. 15-22

---

1. (a) 5700   (b) 10
   (c) 90,000   (d) 320

2. (a) 0.4   (b) 0.02
   (c) 3   (d) 100

3. (a) 3.3, 3.03, 0.3, 0.03
   (b) 0.305, 0.29, 0.05, 0.009

4. (a) 30.06   (b) 73.2

5. (a) 3   (b) 10   (c) 5   (d) 20

---

6. (a) 0.8   (b) 0.1   (c) 2.7   (d) 20.6

7. (a) 590   (b) 2830   (c) 12,100

8. (a) 5700   (b) 13,800   (c) 45,100

9. (a) 4.05   (b) 4.15   (c) −5   (d) −5

10. 4 units = 40
    2 units = 40 ÷ 2 = **20**
    The difference between the two numbers is 20.

11. 1, 7

12. 12, 24

13. 23

14. 11, 13

15. (a) 6097   (b) 364 r2

15. (a) 44   (b) 9

17. (a) >   (b) =   (c) =
    (d) <   (e) >

18. (a) $1\frac{11}{12}$ (b) $\frac{5}{12}$

19. (a) $\frac{2}{3} + \frac{4}{9} = \frac{6}{9} + \frac{4}{9} = 1\frac{1}{9}$

    (b) $\frac{5}{8} + \frac{3}{4} = \frac{5}{8} + \frac{6}{8} = 1\frac{3}{8}$

    (c) $3\frac{3}{10}$ (d) $5\frac{1}{4}$

20. (a) $\frac{3}{5}$ of 15 = 3 x $\frac{15}{5}$ = 3 x 3 = **9**

    (b) $\frac{2}{3}$ of 600 = 2 x $\frac{600}{3}$ = 2 x 200 = **400**

    (c) $\frac{4}{9}$ of 99 = 4 x $\frac{99}{9}$ = 4 x 11 = **44**

21. (a) 4.03   (b) 1.6   (c) 10.85   (d) 5.75

22. (a) $\frac{4}{5}$ (b) $1\frac{1}{4}$ (c) $4\frac{9}{20}$ (d) $6\frac{3}{50}$

23. (a) 10.4, 11.4, 12.7
    (b) 2.4, 3.4, 4.8

24. (a) $\frac{2}{3}$ (b) $\frac{1}{2}$

25. parallelogram

26. C

27.

    1 unit = 215
    (a) There are 4 more units of pencils than pens.
    4 units = 215 x 4 = **860**
    There are 860 more pencils than pens.
    (b) 6 units = 215 x 6 = **1290**
    Or: 215 + 215 + 860 = 1290
    There are 1290 pens and pencils.

28. Adding 2 fifths 5 times gives (2 x 5) fifths:
    $\frac{2}{5}\ell \times 5 = \frac{2 \times 5}{5}\ell = \frac{10}{5}\ell = \mathbf{2\,\ell}$
    She made 2 ℓ of juice.

29. 4 units = $20
    1 unit = $20 ÷ 4 = **$5**

    Or: $\frac{1}{4}$ of $20 = $5
    She had $5 left.

30. $\frac{1}{5}$ of the children cannot swim.
    $\frac{1}{5}$ of 40 = **8**
    8 children cannot swim.

Workbook

Exercise 14, pp. 34-35

1. (a) 74   (b) 10
   (c) 19   (d) 33

2. (a) 47 lb   (b) 3 m
   (c) 1 ℓ   (d) 29 km

3. (a) $3   (b) $11

4. (a) 2 ℓ   (b) 2 ℓ

5. (a) 40   (b) 46
   (c) 6   (d) 6
   (e) 102   (f) 300

Exercise 15, p. 36

1. (a) 4.7
   (b) 8.1

2. (a) 1.5 ℓ
   (b) 20.3 kg
   (c) 9.1 m

3. A: 34.9 kg   B: 41.7 kg   C: 39.8 kg

Review 6, pp. 37-41

1. 92,405

2. thousands

3. 46,495

4. (a) 6000   (b) 42,096   (c) 90,800
   (d) 27,481   (e) −11   (f) −19

5. A: −8   B: 0   C: 10

6. 78,502

7. (a) 0.03   (b) −10

8. 24,519

9. 30, 60

10. (a) $100 - 75 + \underline{48 \div 3}$ (b) $40 + 13 \times \underline{(12 + 6)}$
    $\quad \underline{100 - 75} + 16$ $40 + \underline{13 \times 18}$
    $\quad \underline{25 + 16}$ $40 + 234$
    $\quad \textbf{41}$ $\textbf{274}$

    (c) $1475 - \underline{(18 \times 21)}$ (d) $900 - \underline{(600 - 143)}$
    $\quad \underline{1475 - 378}$ $\underline{900 - 457}$
    $\quad \textbf{1097}$ $\textbf{443}$

11. $\frac{8}{12}$

12. $\frac{7}{12}$

13. 13

14. 3.4

15. A: 1.21   B: 1.28   C: 1.32

16. 4.5, 5

17. $3 \text{ yd} - \frac{5}{6} \text{ yd} = \mathbf{2\frac{1}{6}}$ **yd**

18. 1 unit = 15
    2 units = **30**

19. Adding 2 thirds 15 times gives (2 x 15) thirds:
    $\frac{2}{3} \text{ m} \times 15 = \frac{2 \times 15}{3} \text{ m} = \frac{30}{3} \text{ m} = \textbf{10 m}$

20. 7 units = $35
    1 unit = $35 ÷ 7 = $5
    5 units = $5 x 5 = **$25**

    Or: $\frac{5}{7}$ of $35 = 5 x $\frac{35}{7}$ = 5 x $5 = $25

21. (a) 10 cm   (b) 5 cm   (c) 40 cm   (d) 100 cm²
    (e) rectangle, rhombus, parallelogram, square

22. 125°

23. ∠d

24. 3 units = 1650
    1 unit = 1650 ÷ 3 = 550
    2 units = 550 x 2 = **1100**
    There were 1100 boys.

25. Amount paid in installments = 8 x $95 = $760
    $160 + $760 = **$920**
    She paid $920 altogether.

26.

    3 units = 36
    1 unit = 36 ÷ 3 = 12
    Half of one unit = **6**
    Or: 6 half-units = 36
        1 half-unit = 36 ÷ 6 = 6
    6 girls wear glasses.

27. Perimeter of rectangle: (13 cm + 19 cm) x 2 = 64 cm
    Perimeter of square: 64 cm
    Side of square: 64 cm ÷ 4 = **16 cm**
    Each side if the square is 16 cm long.

Unit 7 – The Four Operations on Decimals

Chapter 1 – Addition and Subtraction

Objectives

♦ Add decimal numbers.
♦ Subtract decimal numbers.
♦ Check the reasonableness of the sum or difference.
♦ Solve word problems involving addition or subtraction of decimal numbers.

Material

♦ Place-value discs (0.001, 0.01, 0.1, 1, 10, and 100)
♦ Number cube labeled with 0.4, 0.5, 0.6, 0.7, 0.8, and 0.9
♦ Mental Math 7-14 (appendix pp. a3-a5)

Notes

Students learned the standard algorithms for addition and subtraction of whole numbers in *Primary Mathematics* 2A, 3A, and 4A. The standard algorithms for addition and subtraction of decimals are the same as those for whole numbers. Numbers are aligned vertically and addition or subtraction is done starting from the smallest place value, that is, from right to left. In the process, the value at any place value may have to be renamed.

In writing the problems vertically, students should take care to align the decimal.

The standard algorithms will be introduced using simple addition or subtraction problems where one or both numbers have only one non-zero digit. Simpler problems give your student a bridge to use the algorithm correctly for problems that cannot be easily solved mentally, such as numbers with more than two non-zero digits. In the exercises you can allow your student to use mental math strategies when possible, if she is capable. However, during the lesson, have her rewrite some problems vertically in order to practice aligning the digits and the decimal correctly.

When discussing the steps in using any of the standard algorithms, use place-value names. For example, in using the standard algorithm for 9.79 + 4.86, we first add 9 hundredths and 6 hundredths to get 15 hundredths, which is the same as 1 tenth 5 hundredths. Do not say, "we first add 9 and 6."

Students learned to round whole numbers to estimate answers to problems in *Primary Mathematics* 4A. In this chapter your student will learn to estimate answers to problems involving decimal numbers. Encourage him to make a habit of estimating answers. Estimation helps reduce errors that may result from putting the decimal point in the wrong place and can help him to determine whether an answer is reasonable. Estimation is particularly useful in multiplication and division where there are more places for potential errors.

When using estimation to determine whether an answer is reasonable we want to round to numbers that allow us to find the estimated answer quickly. This could be to the same place for some students, such as 25.48 + 7.64 = 25 + 8, if they find it easy to mentally add 25 and 8. But they could also round the numbers to 30 + 8, and that is sufficient to check the reasonableness of the actual answer. Any estimates provided in this guide as answers are just possible estimated answers; your student might have a more or less precise estimated answer depending on how she rounded the numbers.

In *Primary Mathematics* 3A, students learned mental addition and subtraction of numbers close to 100, or close to a ten, and applied those strategies to money. For example:

$4.55 + $1.99 = $6.54
   Add $2, and then take off 1¢.

$4.51 − $1.97 = $2.54
   Subtract $2, and then add 3¢.

These strategies will be extended to decimal numbers in this chapter.

Your student will be solving two-step word problems involving adding or subtracting decimals. In *Primary Mathematics* 3A students learned how to use the part-whole and comparison models to solve word problems involving addition and subtraction. The drawing in Task 35 on p. 45 is an example of a part-whole model, and the one in Task 36 on the same page is an example of a comparison model. Modeling is used, but not re-taught here.

The purpose of the model method is to provide a way for students to translate the words in the problem into a visual picture, from which they can see what equations they must use. Even if your student can solve the problems in this chapter without them, understanding these models will be useful to solve more complex word problems in any supplementary books and in *Primary Mathematics* 5A. If this is your first use of the *Primary Mathematics* curriculum, you should provide an introduction to this method from the material in the *Primary Mathematics* 3A textbook or elsewhere, even if your student is able to solve the problems in this chapter without using the models. Use the models during lessons, and have your student draw them when doing the tasks, but use your discretion in requiring them for independent work. If he does have difficulty with the word problems, encourage or require him to draw the models, particularly in correcting problems he gets wrong for other reasons than computation error.

If your student is able to do most of the problems in this chapter without using models, you should consider getting one of the more challenging supplements where she will more likely need to draw models.

Do not get caught up in the process of drawing the models and require your student to follow a set of arbitrary steps in drawing them in such a way that the process becomes more important than the end result (the answer to the problem). Do not establish a set of steps to be applied to all problems. Any such set of steps that works with some problems may not be the best way to approach a different problem, and your student needs experience in being able to use logic and problem-solving skills when using model-drawing, not in applying a set of steps.

Solutions using a model provided in this guide are just suggested solutions to help you with an idea of where to "lead" your student if he is having trouble - other solutions are possible.

For a more advanced student competent with addition, mental math, and place value, you may want to combine lessons, such the first two and then the next two.

(1) Add tenths and hundredths

Discussion

Concept p. 35

This page illustrates adding and subtracting tenths using measurement as the concrete introduction. In this example there is no renaming.

$0.7 + 0.2 = \mathbf{0.9}$
They drank **0.9** liter of milk together.
$0.7 - 0.2 = \mathbf{0.5}$
David drank **0.5** liter more than John.

Task 1, p. 36

1. (a) 0.7 (b) 0.07

This task moves to a more pictorial representation of addition of tenths and hundredths. You can introduce this task more concretely by providing your student with actual place-value discs. Tenths are added to tenths, and hundredths to hundredths, in the same way that ones are added to ones.

Tasks 2-3, p. 36

Copy the problems and discuss their solutions. These tasks illustrate addition of tenths or hundredths when the answer is more than 10 tenths or 10 hundredths. Your student should have no difficulty with the concept of renaming 10 tenths as 1 whole, or 10 hundredths as 1 tenth from the previous unit. If she does have difficulty, use the reinforcement activity below.

Point out that we can write the numbers vertically to keep track of the place value of each digit. When we write the numbers this way, it is important to align the decimal point. That way, tenths are aligned with tenths and hundredths with hundredths (assuming your student uses the same size for each digit; if not, or he has messy handwriting, have him use graph paper or lined paper turned sideways to help with positioning the digits as if on a place-value chart).

Practice

Task 4, p. 37

4. (a) 0.8	(b) 1.3	(c) 1.2
(d) 0.06	(e) 0.1	(f) 0.17

Your student can probably answer these mentally. If you feel it beneficial for her, have her rewrite some of them vertically.

Workbook

Exercise 1, p. 42 (answers p. 45)

Reinforcement

Provide your student with place-value discs and draw a place-value chart. Write an addition problem where tenths are added to tenths or hundredths to hundredths. Have your student pick out the discs, place them on the place-value chart, and replace 10 tenths with 1 whole or 10 hundredths with 1 tenth as needed.

(2) Add 1-place decimal numbers

Discussion

Tasks 5-6, p. 37

5: This task shows that you can split a number into the whole and fractional part and add tenths in the same manner as in the previous lesson. Ask your student for a different mental math strategy. We can also add 0.4 to 6.9 by making the next whole. Tell her that we can use either strategy when simply adding a tenth to a decimal number.

6: Copy the problem and go through the steps for solving it. This task shows the standard algorithm for addition. Review the steps with your student, using place-value names for each digit. Use actual place-value discs if needed. We first add the digits in the smallest place value, which in this case is tenths. Since there are 14 tenths, we rename them as 1 whole and 4 tenths, and write the 1 whole above the ones column to remember to add it to the rest of the ones.

Discuss ways to solve this problem mentally, such as adding the ones first and then the tenths. We are adding 36 tenths and 18 tenths. We can add them the same way as adding 36 ones and 18 ones. We then have to remember that the answer is tenths, not ones, and put in the decimal point accordingly.

5. 7.3
6. 5.4

$$6.9 + 0.4 = 7 + 0.3 = 7.3$$
$$\underset{0.1\ \ 0.3}{\diagup\diagdown}$$

$$3.6 + 1.8 = ?$$
$$36 + 18 = 46 + 8 = 54$$
$$\underset{4\ \ \ 4}{\diagup\diagdown}$$
$$3.6 + 1.8 = 4.6 + 0.8 = 5.4$$
$$\underset{0.4\ \ 0.4}{\diagup\diagdown}$$

Practice

Task 7, p. 37

Allow your student to use either the standard algorithm or mental math, depending on what she needs to work with the most.

7. (a) 8.5	(b) 3.5	(c) 4
(d) 3.3	(e) 6	(f) 6.7

Workbook

Exercise 2, p. 43 (answers p. 45)

Game

Material: Number discs (10's, 1's, and 0.1's). Number cube with 0.4, 0.5, 0.6, 0.7, 0.8, and 0.9.

Procedure: Each player starts with five 1-discs and five 0.1-discs. They write down 5.5. Players take turns throwing the number cube. They collect 0.1-discs according to the number showing face-up on the number cube, trading in ten 0.1-discs for a 1-disc as needed. You can have the player write an addition equation for each throw. The player who collects a 10-disc first wins.

Enrichment

Mental Math 7

(3) Add 2-place decimal numbers

Discussion

Tasks 8-9, p. 38 and Task 11, p. 39

| 8. (a) 1.32   (b) 0.51 |
| 9. 0.61 |
| 11. 6.47 |

8: This task shows that when we add tenths or hundredths to a 2-place decimal we add the tenths to the tenths and the hundredths to the hundredths. You can ask your student for a different mental math strategy for 8(b). We can make the next tenth with either 0.42 or 0.09, taking the hundredths from the other number. We could also, in this case, add a tenth and subtract a hundredth, since 0.09 is one hundredth less than a tenth.

$0.42 + 0.09 = 0.5 + 0.01 = 0.51$
$\qquad \diagup \diagdown$
$\qquad 0.08 \;\; 0.01$

$0.42 + 0.09 = 0.41 + 0.1 = 0.51$
$\diagup \diagdown$
$0.41 \;\; 0.01$

$0.42 + 0.09 = 0.42 + 0.1 - 0.01$
$\qquad\qquad = 0.52 - 0.01$
$\qquad\qquad = 0.51$

9: This task shows the standard algorithm for addition. Review the steps with your student. Use actual place-value discs if needed. We first add the digits in the smallest place value, which in this case is hundredths. Since there are 11 hundredths, we rename them as 1 tenth and 1 hundredth, and write the 1 tenth above the tenths column to remember to add it to the rest of the tenths.

Discuss ways to solve this problem mentally. We are adding 24 hundredths and 37 hundredths. We can add them the same way as adding 24 ones and 37 ones. We then have to remember that the answer is hundredths, not ones, and put in the decimal point accordingly.

$0.24 + 0.37 = ?$
$24 + 37 = 54 + 7 = 61$
$\qquad\qquad\quad \diagup \diagdown$
$\qquad\qquad\quad 6 \;\; 1$
$0.24 + 0.37 = 0.54 + 0.07 = 0.61$
$\qquad\qquad\qquad\quad \diagup \diagdown$
$\qquad\qquad\qquad 0.06 \;\; 0.01$

11: Copy the problem and go through the steps for solving it. In this task, we are adding the equivalent of 3-digit numbers. Go through the steps with your student. When numbers have more digits, it is sometimes easier to rewrite the number vertically and add starting with the digits in the smallest place value, particularly if renaming occurs in more than one place or over several places.

You can ask your student to do another problem on her own, such as 49.79 + 4.86. The steps are exactly the same as with whole numbers.

```
  1 1   1
  4 9 . 7 9
+    4 . 8 6
  5 4 . 6 5
```

Practice

Task 10, p. 38

10. (a) 2.63	(b) 0.96	(c) 1.14
(d) 6.02	(e) 0.4	(f) 1.03
(g) 4.28	(h) 1.18	(i) 1.35
(j) 7.49	(k) 3.06	(l) 4

Workbook

Exercise 3, pp. 44-45 (answers p. 45)

Enrichment

Mental Math 8-9

(4) Estimate sums

Discussion

Tasks 12-13, p. 39

12: Remind your student that we can estimate the answer to a problem by rounding each number that we are adding together. How we round, and to which digit, depends on the situation. If, for example, we wanted to estimate if we had enough money to buy two items that cost $4.80 and $2.37, we might want to round both numbers up, that is, $5 + $3, rather than the second number to the nearest whole number. If we want a quick estimate to see if an exact answer makes sense or has the decimal point in the right place, we could to round to a number with a single non-zero digit. We would round 34.26 + 10.82 to 30 + 10 to give the estimated answer of 40. Sometimes, it is helpful to have a closer estimate. Since it is easy to add 2-digit numbers mentally, the numbers in this example are rounded to the nearest whole number. This gives a closer estimate to the actual value. You can ask your student to find the actual sum, 45.08.

13. (a) 33.12	(b) 7.17

13(a): We could round to the nearest whole number and use mental math to find the estimated value of 33. Or, we could round the first number to 30 and the second to 8, so the estimated value is 38. Rounding both to the nearest whole number gives a more precise estimate, but rounding to one non-zero digit is sufficient to determine if the answer is reasonable with respect to the placement of the decimal point.

$$25.48 + 7.64$$
$$\downarrow \qquad \downarrow$$
$$25 \qquad 8 = 33$$
$$\text{Or} \quad 30 \qquad 8 = 38$$

13(b): Here, we can round both to the nearest whole number to determine if the answer is reasonable. When using estimation to determine if the actual answer is reasonable, we want to round in such a way as to find an estimated answer quickly.

$$4.80 + 2.37$$
$$\downarrow \qquad \downarrow$$
$$5 \qquad 2 = 7$$

Practice

Task 14, p. 39

Workbook

Exercise 4, p. 46 (answers p. 45)

14. (a) 9.76	(b) 6.34	(c) 7.18
(d) 5.92	(e) 9.43	(f) 13.08

(5) Subtract tenths

Discussion

Tasks 15-16, p. 40

15(a): This task shows that we subtract decimals in the same way as we subtract whole numbers. We can subtract 2 tenths from 8 tenths in the same way we can subtract 2 from 8, but they are tenths instead.

15(b-c): To subtract tenths from ones, we need to rename a one as 10 tenths.

16: Copy the problem and go through the steps for solving it. This task shows the standard algorithm for subtraction. Discuss the steps with your student. We do not have enough tenths to subtract 8 tenths, so we rename a one as 10 tenths, giving us 12 tenths. Then we subtract 8 tenths from that.

Discuss ways to solve the problem mentally, such as subtracting 8 tenths from a one and adding the 2 tenths back in, or, since 8 tenths is almost 1, subtracting 1 and adding back in 2 tenths.

Point out that if we are only subtracting tenths, we can ignore any hundredths, and just include that in the answer. Write the problem 4.26 – 0.8 and have your student find the answer.

Practice

Task 17, p. 40

Workbook

Exercise 5, p. 47 (answers p. 45)

Game

Material: Number discs (1's, and 0.1's). Number cube with 0.4, 0.5, 0.6, 0.7, 0.8, and 0.9.

Procedure: Each player starts with five 1-discs and five 0.1-discs. They write down 5.5. Players take turns throwing the number cube. They take away 0.1-discs according to the number showing face-up on the number cube, trading in a 1-disc for ten 0.1-discs when necessary. You can have the player write a subtraction equation for each throw. The player who gets rid of all her discs first (or does not have enough to subtract from) wins.

Enrichment

Mental Math 10

---

15. (a) 0.6   (b) 0.8   (c) 2.8

$4.2 - 0.8 = ?$

$42 - 8 = 2 + 32 = 34$

$/ \ $
2   40

$4.2 - 0.8 = 0.2 + 3.2 = 3.4$

$/ \ $
0.2   4

$4.2 - 0.8 = 4.2 - 1 + 0.2 = 3.4$

$4.26 - 0.8 = 3.4 + 0.06 = 3.46$

$/ \ $
0.06   4.2

$$\begin{array}{r} 3 \\ 4.^{1}2\,6 \\ -\ 0.\,8 \\ \hline 3.\,4\,6 \end{array}$$

---

17. (a) 0.2   (b) 0.2   (c) 0.7
    (d) 0.6   (e) 1.3   (f) 3.1
    (g) 0.6   (h) 4.1   (i) 4.4
    (j) 0.28   (k) 3.55   (l) 4.72

(6) Subtract hundredths

Discussion

Tasks 18-19, p. 41, and Task 21, p. 42

18(a): This task shows that we subtract hundredths from hundredths in the same way as we subtract whole numbers. 8 hundredths – 6 hundredths = 2 hundredths.

18(b): To subtract hundredths from tenths, we need to rename a tenth as 10 hundredths.

18(c): To subtract hundredths from ones, we need to rename a one as 10 tenths, and one of those tenths as 10 hundredths. Point out that 1 is 100 hundredths. We can use mental math strategies for subtracting from a hundred.

19: To subtract tenths and hundredths from 1, we can rename the one as 9 tenths and 10 hundredths, and then subtract the tenths from 9 tenths and the hundredths from 10 hundredths. We use the same strategies for making 100.

21: Copy the problem and go through the steps for solving it. We can also use the standard algorithm to subtract. Point out that 4.2 does not have any hundredths, but we are subtracting hundredths. When we rewrite the problem vertically, we include 0 for the hundredths digit. After renaming a tenth, we have 10 hundredths. When rewriting the problem vertically, we have to be careful to align the digits correctly, not like the incorrect example shown at the right. If we have done so, the decimal point is also aligned.

Write the problem 4.26 – 0.08 vertically and discuss its solution using the standard algorithm. This time, after renaming a tenth, we have 16 hundredths. Point out that we can also use the same mental math strategies we would use for 26 – 8.

Practice

Task 20, p. 41, and Task 22, p. 42

Workbook

Exercise 6, pp. 48-49 (answers p. 45)

Enrichment

Mental Math 11-12

---

18. (a) 0.02   (b) 0.04   (c) 0.94

19. 0.77

$1 - 0.06 = ?$
$100 - 6 = 94$
$1.00 - 0.06 = 0.94$

$1 - 0.23 = ?$
$100 - 23 = 77$
$1.00 - 0.23 = 0.77$

$$\begin{array}{r} 4.2\!\!\!/ \\ -\ 0.0\!\!\!/8 \\ \hline \end{array}$$

$4.26 - 0.08 = ?$

$$\begin{array}{r} \overset{1}{} \\ 4.\,\overset{1}{2}6 \\ -\ 0.08 \\ \hline 4.18 \end{array}$$

$26 - 8 = 6 + 12 = 18$
$\diagup\diagdown$
$6\quad 20$

$4.26 - 0.08 = 4.18$
$\diagup\ |\ \diagdown$
$4\ 0.06\ 0.20$

---

20. (a) 0.07   (b) 0.47   (c) 3.47
    (d) 0.06   (e) 0.26   (f) 2.26
    (g) 0.93   (h) 1.93   (i) 3.91
    (j) 0.55   (k) 2.55   (l) 3.14

22. (a) 3.23   (b) 3.47   (c) 4.16
    (d) 4.74   (e) 6.13   (f) 6.41

(7) Subtract 1-place decimal numbers

Discussion

Task 23, p. 42

> Copy the problem and go through the steps for solving it. As you copy it, explain why you add a decimal point and a 0 to 6 for the tenths place. It helps in keeping track of the place-values; since we are subtracting tenths we will need to rename a one in order to do so. Review the steps with your student, using place-value names for each digit. Use actual place-value discs if needed. There are not enough tenths to subtract 7 from, so we rename 6 ones as 5 ones and 10 tenths. We can then subtract 7 tenths from the 10 tenths.

> Discuss ways we can solve this problem mentally, such as using the same strategies we would use for 60 − 27. We could subtract tens first, and then ones. But instead of the answer being ones, it is tenths; 60 tenths − 27 tenths = 33 tenths = 3.3. Tell your student that if he is unsure of the mental math, he can always rewrite the problem vertically and solve it that way.

23. 3.3

6 − 2.7 = ?
60 − 27 = 33
6.0 − 2.7 = 3.3

Practice

Task 24, p. 42

Workbook

Exercise 7, p. 50 (answers p. 45)

Enrichment

Mental Math 13

24. (a) 3.6   (b) 3.5   (c) 2.7
    (d) 2.5   (e) 2.6   (f) 4.8

(8) Subtract 2-place decimal numbers

Discussion

Tasks 25-26, p. 43

25: Copy the problem and go through the steps for solving it. To subtract the tenths, we need to rename a one as ten tenths. The process is the same as if we were subtracting two 3-digit numbers. However, in copying the problem, it is important to remember to include the decimal point, since we are subtracting ones, tenths and hundredths, not hundreds, tens, and ones.

26: When we rewrite the problem vertically, we can add trailing zeros in order to have the same number of digits after the decimal point. This makes it easier to align the digits correctly. For example, in 26(a), there are hundredths in the first number (7.24), but not the second (3.5), so we add 0 for hundredths (3.50). It is important to remember to include the decimal point when copying the numbers, since they should also align.

25. 1.74		
26. (a) 3.74	(b) 0.31	
3.74	0.31	
(c) 3.73	(d) 2.66	
3.73	2.66	

Practice

Task 27, p. 43

Workbook

Exercise 8, pp. 51-52 (answers p. 45)

27. (a) 0.42	(b) 0.25	(c) 0.88
(d) 3.4	(e) 3.49	(f) 3.55
(g) 0.44	(h) 2.15	(i) 1.62
(j) 1.55	(k) 3.44	(l) 0.95

(9) Estimate sums and differences

Discussion

Task 28, p. 44

This task shows that we can estimate answers for subtraction by rounding the numbers. In this lesson, the purpose for finding an estimate is to check the reasonableness of the answer, particularly with regard to place value. So we want to round the numbers to values that allow quick mental calculation. Rounding both numbers to the nearest one works well here, since it is easy to subtract 8 from 28.

Practice

Tasks 29-30, p. 44

Only the exact answer is given at the right. Your student's estimated answer could vary depending on how he rounds the numbers, but should be close to the actual answer.

29. (a) 15.87 (b) 50.01 (c) 39.57	
30. (a) 2.63 (b) 21.89 (c) 27.54	

Discussion

Tasks 31-33, p. 44

31: 2.99 is 0.01 less than 3, so we can find the answer by adding 3 and then subtracting 0.01.

32: 8.99 is 0.01 less than 9, and 0.99 is 0.01 less than 1. So we can add 9 + 1, and then subtract both extra 0.01's, i.e. 0.02. Or, we could just add 1 to 8.99 and subtract 0.01.

33: 1.99 is 0.01 less than 2. If we subtract 2, we have subtracted 0.01 too much, so we need to add that back in.

31. 7.27
32. 9.98
33. 3.63

Practice

Task 34, p. 44

34. (a) 5.86	(b) 10.8	(c) 9.98
(d) 3.53	(e) 2.04	(f) 4.11

Have your student do a few problems with numbers close to 1 whole but which differ by more than just 0.01, such as those at the right.

$6.29 + 4.98 = 11.29 - 0.02 = 11.27$

$9.97 + 34.2 = 44.2 - 0.03 = 44.17$

$9.45 - 4.95 = 4.45 + 0.05 = 4.5$

Workbook

Exercise 9, p. 53 (answers p. 45)

Enrichment

Mental Math 14

Discussion

Tasks 35-37, pp. 45-46

Discuss these problems with your student. Discuss which parts of the model correspond to the information in the word problem.

35: We are given three parts and asked to find a whole, so the text shows a part-whole model.

36: Since we are told how much longer one ribbon is than the other, the two lengths are being compared, so the text shows a comparison model and one solution where we find the length of the blue ribbon first and then add the two lengths. Since the blue ribbon is the same length as the white ribbon plus another part, we could also combine steps and simply write the expression: 1.85 + 1.85 + 1.4.

37: We are given the whole and two parts, and need to find a missing part.

35.	$11.14
	$11.14
36.	5.1
	5.1 m
37.	$32.95
	$32.95
	$32.95
	$32.95

Practice

You can have your student do several of the word problems from Practice A or B on pages 47 and 48 of the textbook as part of the lesson.

Workbook

Exercise 10, pp. 54-56 (answers p. 45)

(11) Practice

Practice

Use your discretion in requiring your student to draw models for the word problems in these two practices. Any models provided in the solutions are just suggested models. Not all answers will show solutions; likely your student will not need to draw models for all of them.

Practice A, p. 47

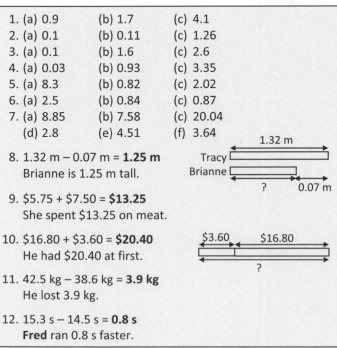

1. (a) 0.9   (b) 1.7   (c) 4.1
2. (a) 0.1   (b) 0.11   (c) 1.26
3. (a) 0.1   (b) 1.6   (c) 2.6
4. (a) 0.03   (b) 0.93   (c) 3.35
5. (a) 8.3   (b) 0.82   (c) 2.02
6. (a) 2.5   (b) 0.84   (c) 0.87
7. (a) 8.85   (b) 7.58   (c) 20.04
   (d) 2.8   (e) 4.51   (f) 3.64

8. 1.32 m − 0.07 m = **1.25 m**
   Brianne is 1.25 m tall.

9. $5.75 + $7.50 = **$13.25**
   She spent $13.25 on meat.

10. $16.80 + $3.60 = **$20.40**
    He had $20.40 at first.

11. 42.5 kg − 38.6 kg = **3.9 kg**
    He lost 3.9 kg.

12. 15.3 s − 14.5 s = **0.8 s**
    Fred ran 0.8 s faster.

Practice B, p. 48

1. (a) 48.68   (b) 19.43   (c) 40.02
2. (a) 28.6   (b) 17.31   (c) 19.98
3. (a) 13.33   (b) 22.23   (c) 4.89
4. (a) 36.65   (b) 11.05   (c) 10.61

5. 1.69 lb + 1.69 lb + 2.51 lb = **5.89 lb**
   The total weight is 5.89 lb.

Reinforcement

Extra Practice, Unit 7, Exercise 1, pp. 115-120

6. 3 ℓ − 0.5 ℓ − 0.25 ℓ = **2.25 ℓ**
   She had 2.25 ℓ of milk left.

Test

Tests, Unit 7, 1A and 1B, pp. 23-26

7.
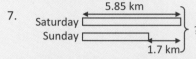

5.85 km + (5.85 km − 1.7 km) = 5.85 km + 4.15 km = **10 km**
He jogged a total of 10 km.

8. 24.8 cm + 12.6 cm + 18.4 cm = **55.8 cm**
   She had 55.8 cm of ribbon at first.

9. $15 − ($4.15 + $6.80) = $15 − $10.95 = **$4.05**
   Lucy spent $4.05

10. $15 − ($4.90 + $7.50) = $15 − $12.40 = **$2.60**
    She received $2.60 change.

Workbook

Exercise 1, p. 42

1. (a) 0.8   (b) 1.2   (c) 0.6
   (d) 1   (e) 1.4
2. (a) 0.06   (b) 0.12   (c) 0.05
   (d) 0.1   (e) 0.11

Exercise 2, p. 43

1. (a) 3.1   (b) 5.4   (c) 10.5   (d) 6.2
2. (a) 5   (b) 8.3   (c) 13.7   (d) 16.3

Exercise 3, pp. 44-45

1. (a) 2.73   (b) 2.55   (c) 5.05   (d) 4.57
   (e) 6.24   (f) 3.88   (g) 2.7   (h) 4.34
2. (a) 0.92   (b) 3.03
   (c) 2.36   (d) 28.28
   (e) 3.62   (f) 9.61
   (g) 17.34   (h) 68.18

Exercise 4, p. 46

1. A: 15 + 30 = 45; 42.9   L: 9 + 12 = 21; 20.51
   H: 42 + 2 = 44; 44.09   C: 70 + 20 = 90; 90
   A: 5 + 6 = 11; 11.36   E: 30 + 40 = 70; 66.9
   W: 20 + 10 = 30; 33.6   I: 54 + 9 = 63; 63
   T: 25 + 3 = 28; 27.35   N: 30 + 60 = 90; 88.75
   R: 60 + 8 = 68; 68.05   G: 78 + 4 = 82; 82
   GREAT WALL OF CHINA

Exercise 5, p. 47

1. (a) 0.6   (b) 0.9   (c) 0.3   (d) 3.9
2. (a) 5.3   (b) 2.6   (c) 3.16   (d) 2.2

Exercise 6, pp. 48-49

1. (a) 0.05   (b) 0.65
   (c) 0.85   (d) 0.92
2. (a) 4.38   (b) 1.48
3. (a) 0.42   (b) 3.24
   (c) 2.78   (d) 6.06
   (e) 2.62   (f) 4.23
   (g) 5.04   (h) 3.91

Exercise 7, p. 50

1. (a) 2.1   (b) 2.7
   (c) 3.6   (d) 1.6
   (e) 2.2   (f) 1.4
   (g) 4.1   (h) 3.6

Exercise 8, pp. 51-52

1. (a) 2.44   (b) 2.55
   (c) 0.07   (d) 8.78
   (e) 3.24   (f) 4.76
   (g) 6.15   (h) 5.43

2. T: 2.35   E: 3.08   H: 0.43   U: 4.65
   R: 4.67   P: 0.78   C: 7.24   S: 1.37
   I: 7.38   G: 4.16   O: 8.96   N: 6.78
   PENGUIN
   OSTRICH

Exercise 9, p. 53

1. (a) 7.24; 7.23; 7.23
   (b) 11.63; 11.58; 11.58
   (c) 1.82; 1.83; 1.83
   (d) 4.05; 4.07; 4.07
2. (a) 9.79   7 + 3 = 10
   (b) 10.64   9 + 2 = 11
3. (a) 4.26   8 − 4 = 4
   (b) 4.58   8 − 3 = 5

Exercise 10, pp. 54-55

1. 5 yd − 2.35 yd = **2.65 yd**
   He used 2.65 yd of wire.

2. 5 kg − 3.6 kg = **1.4 kg**
   He gained 1.4 kg.

3. $36.45 − $2.54 - **$33.91**
   He spent $33.91.

4. $13.50 − ($1.40 + $2.50) = $13.50 − $3.90 = **$9.60**
   She had $9.60 left.

5. $20 − ($12 + $4.50) = $20 − $16.50 = **$3.50**
   She received $3.50 in change.

6.

   $38.90 + $38.90 + $6.50 = **$84.30**
   She spent $84.30.

7.

   1.63 ft − 0.38 ft − 0.25 ft = **1 ft**
   Ribbon B is 1 ft long.

Chapter 2 – Multiplication

Objectives

♦ Multiply decimal numbers by a 1-digit whole number.
♦ Use estimation to check reasonableness of answers.
♦ Solve word problems involving multiplication of decimal numbers.

Material

♦ Place-value discs (0.01, 0.1, 1, 10, and 100)
♦ Place-value chart
♦ Mental Math 15-16 (appendix pp. a5-a6)

Notes

In *Primary Mathematics* 2A, students learned the standard algorithm for multiplying a whole number by a 1-digit whole number. In *Primary Mathematics* 3B, they learned to multiply money in dollars and cents by a 1-digit whole number by converting the money to cents, multiplying, and converting back to dollars and cents.

In this chapter, the standard algorithm will be extended to 1-place and 2-place decimals.

When writing addition or subtraction problems in vertical format, it is important to always align the digits according to their place value. In multiplication, however, we do not align digits for the two factors, since we need to multiply the single digit with the value in each place value. While it is possible to represent 6.14 x 3 in either of the two ways shown on the right, only the second way will be used. In *Primary Mathematics* 5A, students will learn how to multiply a decimal by a whole number greater than 10 or another decimal, without regard to the decimal point when performing the actual multiplication, and then place the decimal point correctly in the final answer.

$$\begin{array}{r} 6.14 \\ \times \quad 3 \\ \hline 18.42 \end{array}$$

$$\begin{array}{r} 6.14 \\ \times \quad\quad 3 \\ \hline 18.42 \end{array}$$

As with addition and subtraction, always use the place values when discussing the process. 0.6 x 5 is "six tenths times five equals 30 tenths or 3 ones," not "six times five is three."

Students will be asked to estimate their answer in order to check if their answer is reasonable. For multiplication, your student can round the decimal to one non-zero digit, e.g. 27.9 x 3 can be estimated by using 30 x 3 = 90 and 0.279 x 3 can be estimated using 0.3 x 3 = 0.9.

In this chapter, your student will be solving 2-step word problems that involve multiplying decimals. In *Primary Mathematics* 3A students learned how to use the part-whole and comparison models to solve word problems involving multiplication. The drawing in Task 18 on p. 55 is an example of a part-whole model, and the one in Task 19 on the same page is an example of a comparison model. Modeling is used, but not re-taught here. Use the models during lessons, and have your student draw them when doing the tasks, but use your discretion in requiring them for independent work. Do not insist your student follow a prescribed set of steps in drawing the models, since unless they learn to solve these problems logically and develop problem-solving skills that don't rely on following a set of steps, they will have difficulty in *Primary Mathematics* 5A.

(1) Multiply tenths and hundredths

Discussion

Concept p. 49

The top half of this page illustrates multiplying tenths using measurement as the concrete introduction. If we have 0.4 liters in 3 separate beakers, then the total amount is 1.2 liters. We have multiplied 4 tenths by 3 to get 12 tenths, which is the same as 1.2.

> 0.4 x 3 = **1.2**
> She drinks **1.2** liters of milk.

The second half of the page shows the same problem with place-value discs, which is a more abstract illustration, and then pictorially with a bar model. Use actual place-value discs to illustrate the concept if needed. Remind your student that if we multiply 4 ones by 3, each one becomes 3 ones, and we have a total of 12 ones, which is 1 ten and 2 ones. In the same way, when we multiply 4 tenths by 3, each tenth becomes 3 tenths, and we have a total of 12 tenths, which is 1 one and 2 tenths. In the bar model, we have 3 equal units, each with a value of 0.4, and the total is 0.4 + 0.4 + 0.4 = 1.2, or 0.4 x 3 = 1.2.

> 4 ones x 3 = 12 ones = 12
> 4 tenths x 3 = 12 tenths = 1.2

Tasks 1-3, pp. 50-51

1: Just as 2 ones x 4 is 8 ones, so 2 tenths x 4 is 8 tenths, and 2 hundredths x 4 is 8 hundredths.

> 1. (a) 0.8 (b) 0.08
> 2. (a) 2.1 (b) 3
> 3. (a) 0.21 (b) 0.3

2-3: After we multiply tenths or hundredths, using the same math facts for multiplying ones, we rename 10 tenths as 1 one and 10 hundredths as 1 tenth. You can show the process with place-value discs if needed. For 2(b) and 3(b), renaming results in the 0 of 30 no longer being in the answer, as it would be with 6 x 5. You can emphasize this by listing answers for the digit in different place values, including hundreds and tens, as shown on the right. Use the names for each place value.

> $\underline{7}00 \times \underline{3} = \underline{21}00$
> $\underline{7}0 \times \underline{3} = \underline{21}0$
> $\underline{7} \times \underline{3} = \underline{21}$
> $0.\underline{7} \times \underline{3} = \underline{2.1}$
> $0.0\underline{7} \times \underline{3} = 0.\underline{21}$
>
> $\underline{6}00 \times \underline{5} = \underline{30}00$
> $\underline{6}0 \times \underline{5} = \underline{30}0$
> $\underline{6} \times \underline{5} = \underline{30}$
> $0.\underline{6} \times \underline{5} = \underline{3.0} = 3$
> $0.0\underline{6} \times \underline{5} = 0.\underline{30} = 0.3$

Practice

Tasks 4-6, p. 51

Your student should be able to do these mentally; it is not necessary to ask him to rewrite the problems vertically here.

> 4. (a) 6   (b) 0.6   (c) 0.06
>   (d) 28   (e) 2.8   (f) 0.28
>   (g) 40   (h) 4   (i) 0.4
> 5. $3.20
> 6. (a) $0.80 (b) $4.20 (c) $7.20

Workbook

Exercise 11, p. 57-58 (answers p. 52)

Reinforcement

Mental Math 15

(2) Multiply 1-place decimals

Discussion

Tasks 7-9, p. 52

7. 12.9
8. 124.2

7: To multiply 4.3 by 3, we need to multiply each digit by 3. 4 ones x 3 is 12 ones and 3 tenths x 3 is 9 tenths, so we have 12 ones and 9 tenths. Renaming ones gives 1 ten 2 ones.

```
4.3 x 3
 /\
4  0.3
4 x 3 = 12
0.3 x 3 = 0.9
12 + 0.9 = 12.9
```

8: This task shows the result of multiplying the digit in each place by 6. Since 7 tenths x 6 = 42 tenths, 40 tenths need to be renamed as 4 ones. Remind your student that we can do the multiplication in steps, starting with the lowest place value. The process is exactly the same as we used with whole numbers. Go through the steps for the multiplication algorithm. If necessary, show the same steps with the place-value discs, i.e. first multiply the seven tenths and rename the answer before multiplying the 2 tens, as was done in *Primary Mathematics* 3A.

```
        4
    2 0 . 7
  x       6
      . 2        (0.7 x 6 = 4.2)
        ↓
        4
    2 0 . 7
  x       6
    4 . 2        (0 x 6 + 4 = 4)
        ↓
        4
    2 0 . 7
  x       6
  1 2 4 . 2      (20 x 6 = 120)
```

9: To estimate the answer to a multiplication problem, we generally round both factors to a number that has only one non-zero digit (unless rounding differently gives a problem that we already know the answer to, e.g. 25.8 x 4; 25 x 4 = 100 is a good estimation easy to calculate). Make sure your student understands that using estimation allows her to do a quick check on whether the answer is reasonable with regard to where the decimal point is.

Practice

Task 10, p. 52

Workbook

Exercise 12, p. 59 (answers p. 52)

10. (a) 6 x 2 = 12	(b) 4 x 5 = 20	(c) 8 x 30 = 240
11.8	19.5	260.8
(d) 20 x 2 = 40	(e) 3 x 30 = 90	(f) 100 x 4 = 400
37	80.4	520.8

(3) Multiply 2-place decimals

Discussion

Tasks 11-15, pp. 53-54

11-12: Rewrite the problems and discuss each step.

13: Have your student rewrite the problems and work through the steps.

14: Doing a quick estimation helps us know if we placed the decimal point correctly. From the estimate, we would know that 615.6, which would result from not placing the decimal point correctly in the answer, is not a reasonable answer.

15: Point out that the girl's method for estimating the answer is just as valid as the boy's method. It does not give a specific value, but it does tell her that the answer will be just under half of 8. So 33.6 is not a reasonable answer, but 3.36 is. Estimations do not tell us if the answer is correct, just if it is reasonable.

11. 0.75	
12. 9.06	
13. (a) 414.09	(b) 180.81
414.09	180.81

Practice

Tasks 16-17, p. 54

Workbook

Exercise 13, pp. 60-61 (answers p. 52)

Enrichment

Mental Math 16

16. (a) 0.3 x 4 = 1.2	(b) 3 x 4 = 12	(c) 0.5 x 4 = 2
1.04	12.48	1.8
(d) 5 x 8 = 40	(e) 30 x 3 = 90	(f) 50 x 6 = 300
36.16	102.06	289.56
(g) 5 x 40 = 200	(h) 8 x 60 = 480	(i) 30 x 7 = 210
180.75	442	208.11
17. (a) $8.20	(b) $117.00	(c) $292.05

(4) Word problems

Discussion

Tasks 18-21, pp. 55-56

Discuss these problems with your student.

18: Discuss which part of the drawing goes with each piece of information in the question. We are asked to find a whole, and are given the value of one unit, so the text shows a part-whole model.

19: Since we are told how much more Susan saved than Rachel, the amounts both girls saved are being compared, and so the text shows a comparison model. The amount Rachel saved is the unit for which we are given the value. After your student answers the problem, by way of review, you can ask him to also find the difference between what Susan and Rachel saved (1 unit x 3 = $60.15) and the total amount they saved (1 unit x 5 = $100.25).

20-21: You may want to ask your student to draw models for these. There are suggested models at the right; a combination part-whole model for multiplication/division and for addition/subtraction could be used to show all the information.

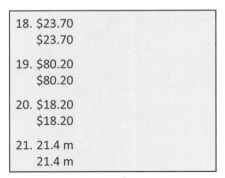

18. $23.70
    $23.70

19. $80.20
    $80.20

20. $18.20
    $18.20

21. 21.4 m
    21.4 m

Practice

You can have your student do several of the word problems from Practice C on p. 57 as part of the lesson.

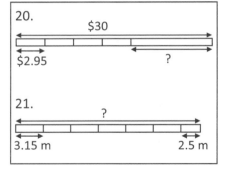

20.
  $30
  $2.95   ?

21.
  ?
  3.15 m   2.5 m

Workbook

Exercise 14, pp. 62-64 (answers p. 52)

Reinforcement

Extra Practice, Unit 7, Exercise 2, pp. 121-124

(5) Practice

Practice

Practice C, p. 57

Test

Tests, Unit 7, 2A and 2B, pp. 27-32

Enrichment

See if your student can solve the following problem. There is more than one method to solve the problem. One method is suggested below.

⇒ If Amy gives $3.65 to Zoe, she will have three times as much money as Zoe. If Amy gives $7.20 to Zoe, she will have twice as much money as Zoe. How much money does Amy have?

In both cases, the total amount stays the same. So draw two bars that are the same for each case.

In the first case, Amy ends up with 3 times as much as Zoe, so there are 4 total units (3 for Amy, one for Zoe). In the second case, there are 3 equal units.

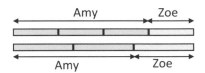

In order to compare the two bars, subdivide them so that there are equal units. Now, we can see that the difference in what she gave Zoe in the two cases is 1 of these smaller units

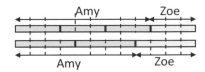

1. (a) 8.6   (b) 11   (c) 2.28
2. (a) 6.23   (b) 4.52   (c) 4.14
3. (a) 5.3   (b) 3.5   (c) 1.85
4. (a) 1.52   (b) 3.17   (c) 2.85
5. (a) 3.6   (b) 5.6   (c) 1.86
6. (a) 1.35   (b) 6   (c) 19.3
7. (a) 18; 19.2 (b) 6; 7.44 (c) 20; 20.45
8. 1.5 m – 1.39 m = **0.11 m**
   The difference between the two results is 0.11 m.

9.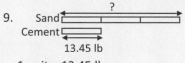
   1 unit = 13.45 lb
   3 units = 13.45 lb x 3 = **40.35 lb**
   He used 40.35 lb of sand.

10. 1 packet spice: $0.85
    4 packets spice: $0.85 x 4 = $3.40
    $3.40 + $3.75 = **$7.15**
    She spent $7.15.

11. (1.46 ℓ + 0.8 ℓ) – 0.96 ℓ = 2.26 ℓ – 0.96 ℓ = **1.3 ℓ**
    He had 1.3 ℓ of gray paint left.

12.
    1 unit = $2.35
    5 units = $2.35 x 5 = $11.75
    $20 – $11.75 = **$8.25**
    She received $8.25 change.

1 unit = $7.20 – $3.65 = $3.55
Using the first case, Amy ends up with 9 units.
9 units = $3.55 x 9 = $31.95
She started with $3.65 more.
$31.95 + $3.65 = $35.60
Amy had $35.60.

Workbook

Exercise 11, pp. 57-58

1. (a) 0.8
   (b) 1.8
   (c) 1.4
   (d) 3.6
   (e) 3
   (f) 5.6
   (g) 2.7
   (h) 4

2. (a) 0.06
   (b) 0.28
   (c) 0.18
   (d) 0.35
   (e) 0.3
   (f) 0.72
   (g) 0.12
   (h) 0.48

Exercise 12, p. 59

1. (a) 8.6   (b) 19.2
   (c) 16.8   (d) 42.3

2. (a) 27.6   7 x 4 = 28   (b) 38.5   7 x 6 = 42
   (c) 132.5   30 x 5 = 150   (d) 244.8   8 x 30 = 240

Exercise 13, pp. 60-61

1. (a) 1.66   (b) 0.72
   (c) 15.78   (d) 27

2. (a) 42.18   7 x 6 = 42   (b) 45.12   8 x 6 = 48
   (c) 579.46   80 x 7 = 560   (d) 582.48   9 x 60 = 540

3. L: 0.96   H: 81.2   E: 0.21   Y: 14.73
   T: 32.25   E: 561   P: 726.3   E: 64.44
   N: 36.45   D: 3265.6   H: 28.94   E: 78.48
   HELP THE NEEDY

Exercise 14, pp. 62-64

1. 1 unit = 1.25 yd
   3 units = 1.25 yd x 3 = **3.75 yd**
   The total length is 3.75 yd.

2. 1 unit = 5.7 ℓ
   5 units = 5.7 ℓ x 5 = **28.5 ℓ**
   The capacity of the fish tank is 28.5 ℓ.

3. 1 unit = $2.50
   6 units = $2.50 x 6 = **$15**
   He saved $15 altogether.

4. 1 can of chocolates: $6.90
   2 packs of crackers: $1.45 x 2 = $2.90
   Total: $9.80

   2 bags of nuts: $3.75 x 2 = $7.50
   2 bottles of sauce: $0.95 x 2 = $1.90
   Total: $9.40

   1 bath towel: $9.95
   4 face towels: $1.20 x 4 = $4.80
   Total: $14.75

   3 dolls: $8 x 3 $24.00
   1 teddy bear: $16.50
   Total: $40.50

5.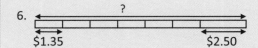

   5 m − (0.85 m x 2) = 5 m − 1.7 m = **3.3 m**
   She had 3.3 m of material left.

6.

   ($1.35 x 6) + $2.50 = $8.10 + $2.50 = **$10.60**
   She had $10.60 at first.

Chapter 3 – Division

Objectives

♦ Divide decimal numbers by a 1-digit whole number.
♦ Use estimation to check reasonableness of answers.
♦ Round the quotient to 1 decimal place.
♦ Solve word problems involving division of decimal numbers.

Material

♦ Place-value discs (0.01, 0.1, 1, 10, and 100)
♦ Mental Math 17-18 (appendix pp. a6-a7)

Notes

In *Primary Mathematics* 3A, students learned the standard algorithm for dividing a whole number by a 1-digit whole number. In *Primary Mathematics* 3B, they divided dollars and cents in the decimal notation by a 1-digit whole number by converting the dollars and cents to cents only, dividing, and then converting back to dollars and cents.

In this chapter, the formal algorithm is extended to dividing a decimal number by a whole number. In *Primary Mathematics* 5A, students will learn how to divide by a decimal.

The dividend is the quantity to be divided, and the divisor is the quantity by which another quantity is to be divided. Because it is far more important for your student to learn division than to learn the formal names of the process, these terms are not used in the course material itself, and will be used only in the guide to avoid having to use terms such as "number being divided."

$$\text{dividend} \div \text{divisor} = \text{quotient}$$

$$\text{divisor}\overline{)\text{dividend}}^{\text{quotient}}$$

Your student will be solving many problems in this section by using mental calculations (i.e., problems where the quotient has one non-zero digit, such as $3.6 \div 6 = 0.6$). He should be thoroughly familiar with the division facts for whole numbers.

When a whole number is divided by another whole number, the answer can be in the form of a whole number quotient with a remainder ($29 \div 4 = 7$ R 1), a fraction or mixed number ($29 \div 4 = 7\frac{1}{4}$), or a decimal ($29 \div 4 = 7.25$). Students have learned to find an answer in the form of a whole number quotient and remainder or a mixed number in earlier levels. In this chapter, your student will learn to find a decimal quotient and to round decimal quotients to 1 decimal place.

Your student will be finding an estimate to check the reasonableness of the actual answer. In division, rather than rounding the dividend to the nearest whole number or ten or other multiple of ten, we round to the closest multiple of the the divisor. For example, to estimate the answer for the division problem $31.2 \div 8$, it would not be any easier to divide 30 by 8 than it is to divide 31.2 by 8, so it is not helpful to round the dividend to the nearest ten in this case. 8 x 3 is 24 and 8 x 4 is 32; 31.2 is closer to 32 than 24, so $31.2 \div 8$ is about 4.

(1) Divide decimals using division facts

Discussion

Concept p. 58

The top half of this page illustrates dividing tenths using measurement as the concrete introduction. If we divide 0.9 liters into 3 equal parts, each part has 0.3 liters.

The second half of the page shows the same problem with place-value discs, and then pictorially with a bar model. Use actual place-value discs to illustrate the concept if needed. Point out that we can rewrite the problem vertically, and solve by finding what number times 3 is 0.9. 0.3 x 3 is 0.9, and there is no remainder, so the quotient is 0.3.

Tasks 1-3, pp. 59-60

1: We can divide 6 tenths or 6 hundredths by 2 in the same way as we divide 6 ones by 2, but the answer is in tenths or hundredths.

2: We divide decimals the same way as we divide whole numbers when using the division algorithm. We start by trying to divide the number in the highest place, which in this case is 1. Since 1 whole cannot be divided by 3, we have to rename it as 10 tenths and add that to the 8 tenths we already have. We can then divide 18 tenths by 3. We use the division fact $18 \div 3 = 6$ to find the quotient. The quotient is 6 tenths. Similarly, to divide 2 by 4, we need to rename 2 as 20 tenths and divide that by 4. Show how you would do the problem vertically by adding a 0 on after the decimal point and writing the answer above the line in the correct place, which is above the 0.

3: We can follow the same process with hundredths, renaming tenths as hundredths as necessary in order to divide.

Practice

Tasks 4-6, p. 60

Your student should be able to do these mentally.

Workbook

Exercise 15, p. 65-66 (answers p. 64)

Reinforcement

Mental Math 17

$0.9 \div 3 = \mathbf{0.3}$
There was **0.3** liter of water in each beaker.

---

1. (a) 0.3   (b) 0.03

2. (a) 0.6   (b) 0.5

3. (a) 0.06   (b) 0.05

---

6 ones ÷ 2 = 3 ones
6 tenths ÷ 2 = 3 tenths
6 hundredths ÷ 2 = 3 hundredths

$$\begin{array}{r} 0.5 \\ 4\overline{)2.0} \end{array}$$

$$\begin{array}{r} 0.05 \\ 4\overline{)0.20} \end{array}$$

---

4. (a) 2   (b) 0.2   (c) 0.02
    (d) 5   (e) 0.5   (f) 0.05
    (g) 6   (h) 0.6   (i) 0.06

5. $0.70

6. (a) $0.30   (b) $0.30   (c) $0.60

(2) Divide hundredths

Discussion

Task 7, p. 61

7. 0.37

Rewrite the problem and discuss the steps used in the division algorithm, as shown in the textbook. They are the same steps as used with whole numbers, only now there are other place values less than ones. The picture of the place-value discs shows that 7 tenths are divided by 2, with 3 tenths in each group, and the remaining tenth, which cannot be divided by 2, is renamed as 10 hundredths, so that there are 14 hundredths to divide by 2.

Discuss some additional problems with your student if needed, either from the next 3 tasks, or ones you make up. This section involves only 2-place decimal numbers less than 1, so only tenths might have to be renamed, and the answer does not go beyond 2 decimal places (no remainder after hundredths are divided).

Practice

Tasks 8-10, p. 61

Workbook

Exercise 16, p. 67 (answers p. 64)

8. (a) 0.13	(b) 0.21	(c) 0.11
(d) 0.17	(e) 0.15	(f) 0.16
9. $0.15		
10. (a) $0.15	(b) $0.15	(c) $0.19

(3) Divide decimals

Discussion

Tasks 11-13, p. 62

11: This task extends division of decimals to a decimal number with 3 non-zero digits where renaming may occur for both the ones and the tenths. Rewrite the problem and discuss the steps used in the division algorithm.

Do some additional problems with your student as needed, either from Tasks 14-16, or ones you make up, where the answer does not go beyond 2 decimal places. You can do one where there is a 0 in the tenths place, such as 7.02 ÷ 3 shown at the right, or one where a 0 is needed in the quotient, such as 30.35 ÷ 5.

12: Point out that to estimate the answer in division, rather than rounding to a number with only one non-zero digit, as we did in multiplication, we round to the closest multiple of the number we are dividing by. 8 x 3 is 24 and 8 x 4 is 32; 31.2 is closer to 32 than 24, so 31.2 ÷ 8 is about 4. You can ask your student to find the exact answer (3.9).

13: Here, the girl is rounding to 5.4 using the multiplication fact 6 x 9 for 6 x 0.9. We could also round 5.4 to 6 to get an estimate of 1. The actual answer is 0.88, and either estimate will tell us that the answer is less than 1, and that an answer of 8.8 or 0.088 would not be reasonable. Ask your student if she can predict which types of problems will give an answer less than 1. Any time the number we are dividing is smaller than the number we are dividing by, the answer will be less than 1.

> 11. 1.45

> ```
>      2
> 3 )7.02    Divide ones.
>    6
>    1
>    ↓
> 
>      2.3
> 3 )7.02    Divide tenths.
>    6
>    1 0
>      9
>      1
>      ↓
> 
>      2.34
> 3 )7.02    Divide hundredths.
>    6
>    1 0
>      9
>      12
>      12
>       0
> ```

Practice

Tasks 14-16, pp. 62-63

Workbook

Exercise 17, problems 1-2, pp. 68-69 (answers p. 64)

> 14. (a) 0.9 ÷ 3 = 0.3   (b) 7.2 ÷ 8 = 0.9   (c) 45 ÷ 9 = 5
>   Or: less than 1   Or: 8 ÷ 8 = 1   5.15
>   0.27   0.89
>
>   (d) 3 ÷ 3 = 1   (e) 4 ÷ 4 = 1   (f) 12 ÷ 6 = 2
>   1.32   1.03   2.43
>
> 15. $2.30
>
> 16. (a) $1.55   (b) $2.75   (c) $3.45
>   (d) $0.71   (e) $0.72   (f) $0.56

(4) Add place values to divide

Discussion

Task 17, p. 63

17. (a) 1.25
    (b) 1.35

17(a): This task shows that when we divide whole numbers and get a remainder, rather than stopping the division there, we can rename the remainder as tenths and continue the division problem, adding 0's in the tenths and hundredths place. Rewrite the problems and discuss the steps. Ask your student if he remembers another way we could express the remainder in the first step. We could express it as a fraction, so the answer would be expressed as a mixed number, $1\frac{1}{4}$. If we continue the division we get 1.25, which is equivalent to $1\frac{1}{4}$.

17(b): This problem shows that after dividing the tenths, if there is a remainder, we can continue the division problem by renaming the remainder tenths as hundredths.

18. (a) 6.08   (b) 1.5
        6.08   1.5

19. (a) 1.6   (b) 2.5   (c) 2.75
    (d) 0.45   (e) 0.34   (f) 4.25

Practice

Tasks 18-19, p. 64

Workbook

Exercise 17, problem 3, p. 70 (answers p. 64)

Exercise 18, pp. 71-72 (answers p. 64)

(5) Round the quotient to 1 decimal place

Discussion

Tasks 20-21, p. 64

20: In the previous lesson, 0's were added when there was a remainder in the ones or tenths place in order to carry out the division to the hundredths place, but in all the previous problems your student has had, there has been no remainder after hundredths were divided. In this problem, there is a remainder of one hundredth.

> 20. 2.3
>   2.3
>
> 21. 19.6

Copy the problem and have your student work it out. She could carry the division out to the thousandths place. Point out that if she continued with the division, adding places each time, she will continue to get 1 as a remainder in that place value. With some division problems, we never get a remainder of 0 in any place. Rather than give the answer as a fraction or with a remainder, we round to a given decimal place, in this case, tenths.

Tell your student that often we only need an answer that is accurate to a certain place value. For example, decimal numbers are primarily used in measurements, and often we cannot measure more accurately than one or two decimal places. If we bought something 7 meters long and wanted to cut it into three equal pieces, we could not measure the pieces more accurately than at best to the nearest centimeter. So we would round the answer to two decimal places and measure each piece to about 2.33 m, or 2 m 33 cm. In other cases, we might want to have our answer only to the nearest tenth. $7 \div 3 = 2.3$ to 1 decimal place.

21: Make sure your student understands that in order to round the answer to 1 decimal place, we need to continue the division to two decimal places to determine whether we need to round the digit in the tenths place up or down. The steps are shown at the right, with the answer to two decimal points and a remainder of 2 hundredths.

```
        19.62
    4) 78.50
       4
       ‾‾
       38
       36
       ‾‾
        2 5
        2 4
        ‾‾‾
         10
          8
          ‾
          2
```

Practice

Task 22, p. 64

Workbook

Exercise 19, p. 73 (answers p. 64)

> 22. (a) 0.2   (b) 0.6   (c) 0.6
>   (d) 0.2   (e) 0.4   (f) 5.4

Enrichment

You can ask your student to carry out some of the divisions in Task 22 to more places to see if he gets a terminating decimal (eventually no remainder) or a non-terminating decimal (always a remainder) and what repeating pattern the digits in the quotient have when the quotient is a non-terminating decimal.

> 22. (a) 0.1<u>6</u>6...
>   (b) 0.<u>571428</u>5...
>   (c) 0.<u>5</u>5...
>   (d) 0.225
>   (e) 0.41<u>6</u>6...
>   (f) 5.4375

Ask your student to evaluate the following expressions.

1 ÷ 9	(0.111…)
2 ÷ 9	(0.222…)
3 ÷ 9	(0.333…)
4 ÷ 9	(0.444…)
5 ÷ 9	(0.555…)
6 ÷ 9	(0.666…)
7 ÷ 9	(0.777…)
8 ÷ 9	(0.888…)

Then ask her what 9 ÷ 9 would be if we continued the pattern (0.9999…). Ask her if she thinks that 0.9999… with the 9's extended forever to smaller and smaller decimal places is the same as 1. (It is not necessary to provide a definitive answer at this time. Mathematically, 0.9999… is really 1. If you multiply both sides of $\frac{1}{3}$ = 0.3333… by 3, you get 1 = 0.9999…).

Ask your student to do the divisions shown below, carrying out each division until he sees the pattern (142857) repeating in the decimal.

1 ÷ 7	(0.142857142857…)
2 ÷ 7	(0.285714285714…)
3 ÷ 7	(0.428571428571…)
4 ÷ 7	(0.571428571428…)
5 ÷ 7	(0.714285714285…)
6 ÷ 7	(0.857142857142…)

See if she realizes that each of these has the repeating pattern, but the pattern starts at different places in the decimal. Once she see this, she only needs to find the first digit where it starts:

At 7 ÷ 7 the pattern breaks down since 7 ÷ 7 = 1. If your student is interested, he can continue and discover that the pattern is consistent with the remainder (e.g., 8 ÷ 7 has a remainder of 1 and the pattern after the decimal is the same as 1 ÷ 7):

8 ÷ 7	(1.142857142857…)
9 ÷ 7	(1.285714285714…)
10 ÷ 7	(1.428571428571…)

(6) Solve word problems

Discussion

Tasks 23-26, pp. 65-66

Discuss these problems with your student.

23: The drawing helps us put the information into a picture and come up with a method to solve the problem. We are given a whole and equal parts and have to find the value in two equal parts. So this involves both division (to find the value of one part) and multiplication (to then find the value of two parts). Point out that in many problems that are complex enough where a diagram like this is useful, we will often have to first find the value of 1 unit before being able to find the answer to the problem.

> 23. 5 units = $8
>   1 unit = $8 ÷ 5 = $1.60
>   2 units = $1.60 x 2 = **$3.20**
>
> 24. 3 units = $5.40
>   1 unit = $5.40 ÷ 3 = $1.80
>   2 units = $5.40 – $1.80 = **$3.60**
>   (Or: 2 units = $1.80 x 2 = $3.60)
>   **$3.60**
>
> 25. 5 units = 5 gal – 0.25 = 4.75 gal
>   1 unit = 4.75 gal ÷ 5 = **0.95 gal**
>   **0.95** gal
>
> 26. 5.4 ÷ 5 = **1.08**
>   **1.08** kg

24: Since we are told how many times more Taylor has than Bonita, two quantities are being compared, so we can draw a comparison model. Again, to answer the problem we first need to find the value of 1 unit.

25: This model shows a combination of a part-whole model for addition and subtraction (unequal units, one unit is the amount in the bottles and the other is the left-over amount) and a part-whole model for multiplication (the amount in the bottles is divided into 5 equal units). We have to find the value of one part by subtraction before we can find the value of 1 unit using division.

26: Ask your student why the solution shows multiplication as the first step. She may not need a model for this problem. If she does, one possibility is to draw two bars of equal length to represent the flour in the bags and in the cake. One bar has 4 units and the other 5.

You may want to show an alternative solution, since in more complex problems your student will encounter later he may have to make equal units between two equal bars such as this. If the unit in the bar with 4 equal units is divided into fifths, and that in the bar with 5 equal units is divided into fourths, there are now equal sized units in both bars. Each bag of flour is now represented by 5 units and each cake by 4 units.

> 26. (alternative solution)
>
>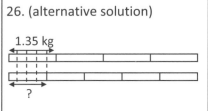
>
> 5 units = 1.35 kg
> 1 unit = 1.35 kg ÷ 5 = 0.27 kg
> 1.35 kg – 0.27 kg = 1.08 kg
> She used 1.08 kg of flour for each cake.

Practice

You can have your student do some of the word problems from the practices on the next three pages in the textbook as part of the lesson.

Workbook

Exercise 20, pp. 74-76 (answers p. 64)

(7) Practice

Practice

Practice D, p. 67

Reinforcement

Extra Practice, Unit 7, Exercise 3, pp. 125-130

Enrichment

Mental Math 18

See if your student can solve the following problem.

⇒ 5 zucchini cost as much as 3 avocados. If 3 zucchini and 6 avocados cost $12.09, how much more does each avocado cost than each zucchini? (All avocados cost the same and all zucchini cost the same.)

⇒ Start with some diagrams:

Zucchini
Avocado

$12.09

We can solve this problem using replacement. In this case, we can replace each set of 3 avocados with 5 zucchini. Then we can find the cost of one zucchini. Since 5 zucchini cost the same a 3 avocados, we can then find the cost of 1 avocado.

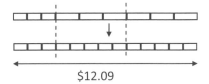

$12.09

1. (a) 32.8 (b) 15.87 (c) 26.32

2. (a) $0.90 (b) $16.20 (c) $30.60

3. (a) 3.2 (b) 0.42 (c) 1.36

4. (a) $0.15 (b) $0.60 (c) $0.85

5. (a) 36; 39.24 (b) 30; 31.5 (c) 14; 13.58

6. (a) 3; 2.95 (b) 4; 3.99 (c) 9; 8.76

7. 4 bottles: 6 qt
Amount in each bottle = 6 qt ÷ 4 = **1.5 qt**
There were 1.5 qt in each bottle.

8. 1 liter: 1.25 kg
6 liters: 1.25 kg x 6 = **7.5 kg**
6 liters of gas weigh 7.5 kg.

9. 5 pieces: 6.75 yd
1 piece: 6.75 yd ÷ 5 = **1.35 yd**
Each piece is 1.35 yd long.

10. 1 pot hanger: $3 + $1.40 = $4.40
4 pot hangers: $4.40 x 4 = **$17.60**
It will cost him $17.60 to make 4 pot holders.

11.

2.34 kg

?   0.06 kg

(2.34 kg – 0.06 kg) ÷ 6 = 2.28 kg ÷ 6 = **0.38 kg**
One bar weighs 0.38 kg.

12.

?

Book
Comic   } $8.25

3 units = $8.25
1 unit = $8.25 ÷ 3 = $2.75
2 units = $2.75 x 2 = **$5.50**
or $8.25 – $2.75 = $5.50
The book costs $5.50.

13 zucchini = $12.09
1 zucchini = $12.09 ÷ 13 = $0.93
5 zucchini = $0.93 x 5 = $4.65
3 avocados = $4.65
1 avocado = $4.65 ÷ 3 = $1.55
$1.55 – $0.93 = $0.62

An avocado costs $0.62 more than a zucchini.

(8) Practice

Practice

Practice E, p. 68

Enrichment

See if your student can solve the following problem.

⇒ 2 avocados and 6 zucchini together cost $8.60. 3 avocados and 3 zucchini together cost $7.80. How much more does each avocado cost than each zucchini? (All avocados cost the same and all zucchini cost the same.)

Start with a diagram:

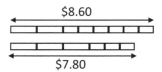

In this problem, we cannot directly replace zucchini with avocado to get one type of unit. One bar is longer than the other, but subtracting one from the other does not give equal units. Rearranging the units does not help either. One way to approach a problem like this is to determine what information we can get out of it.

In the first case, there are an even number of each vegetable. So we can halve the number, and the total cost. Similarly, in the second case, we have the same number of avocado and zucchini. So we can find the cost of 1 avocado and 1 zucchini.

1. 1 can: 5.5 ℓ
   8 cans: 5.5 ℓ x 8 = **44 ℓ**
   He used 44 ℓ of paint.

2. Mrs. Wells = 4 daughters = 47.6 kg
   1 daughter = 47.6 kg ÷ 4 = **11.9 kg**
   Her daughter weighs 11.9 kg.

3. Doll: $4.95
   Robot = 3 dolls: $4.95 x 3 = **$14.85**
   The robot costs $14.85.

4. 3 girls paid: $17.40
   1 girl paid: $17.40 ÷ 3 = **$5.80**
   Each girl paid $5.80.

5. 1 storybook: $2.80
   5 storybooks: $2.80 x 5 = $14
   Change: $20 − $14 = **$6**
   He received $6 change.

6.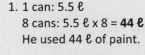
   ($50 − $15.25) ÷ 5 = $34.75 ÷ 5 = **$6.95**
   1 m of cloth cost $6.95.

7. Saved in 4 days: $4.60 x 4 = $18.40
   Saved last day: $25 − $18.40 = **$6.60**
   She saved $6.60 the last day.

8. Cost of tea: $0.65 x 3 = $1.95
   Cost of juice: $4.40 − $1.95 = **$2.45**
   The glass of orange juice cost $2.45.

2 avocados and 6 zucchini: $8.60
1 avocado and 3 zucchini: $8.60 ÷ 2 = $4.30

3 avocados and 3 zucchini: $7.80
1 avocado and 1 zucchini: $7.80 ÷ 3 = $2.60

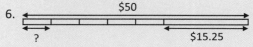

Now that we have an equal number of avocados in both cases, the difference is now equal units of zucchini.

2 zucchini: $4.30 − $2.60 = $1.70
1 zucchini: $1.70 ÷ 2 = $0.85
1 avocado: $2.60 − $0.85 = $1.75
$1.75 − $0.85 = $0.90

An avocado costs $0.90 more than a zucchini.

(9) Practice

Practice

Practice F, p. 69

Test

Tests, Unit 7, 3A and 3B, pp. 33-36

Enrichment

See if your student can solve the following problem.

⇒ 2 avocados and 5 zucchini together cost $7.95. 4 avocados and 3 zucchini together cost $10.65. How much more does each avocado cost than each zucchini? (All avocados cost the same and all zucchini cost the same.)

Start with a diagram:

$7.95

$10.65

In this problem, we cannot take a fraction of the total number and get whole amounts of avocados and zucchini, as in the enrichment problem on the previous page of this guide. But if we can get the same number of avocados or zucchini in both bars, then the difference will be equal units, as in the previous problem. See if your student can come up with a way to manipulate one or both bars to get equal units of avocados or zucchini. One way to do this is to notice that the second bar has 4 avocados and the first has two. So if we double the amount in the first case, we will have 4 avocados and 10 zucchini. The number of avocados is now the same for both cases. The units now can be lined up, and the difference between the bars is a whole number of units of zucchini.

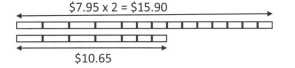

$7.95 x 2 = $15.90

$10.65

1. 1 plate: $2.50
   3 plates: $2.50 x 3 = **$7.50**
   They paid $7.50 altogether.

2. 4 pillow cases: 6.6 m
   1 pillow case: 6.6 m ÷ 4 = **1.65 m**
   She used 1.65 m of lace for each.

3. 1.8 kg + (2.05 kg x 3) = 1.8 kg + 6.15 kg = **7.95 kg**
   The 4 packages weighed 7.95 kg.

4. $5.65 + ($1.45 x 6) = $5.65 + $8.70 = **$14.35**
   She spent $14.35.

5. 2 people: $6.70
   1 person: $6.70 ÷ 2 = $3.35
   $15.35 − $3.35 = **$12**
   Alice had $12 left.

6.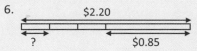
   $2.20

   ? $0.85

   ($2.20 − $0.85) ÷ 3 = $1.35 ÷ 3 = **$0.45**
   Each pencil cost $0.45.

7.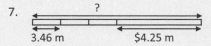
   ?

   3.46 m $4.25 m

   (3.46 m x 3) + 4.25 m = 10.38 m + 4.25 m = **14.63 m**
   She used 14.63 m of material.

8. 4 apples: $2.20
   1 apple: $2.20 ÷ 4 = $0.55
   $0.60 - $0.55 = **$0.05**
   An apple is $0.05 cheaper at the sale.

7 zucchini: $15.90 − $10.65 = $5.25
1 zucchini: $5.25 ÷ 7 = $0.75
3 zucchini: $0.75 x 3 = $2.25
4 avocados: $10.65 − $2.25 = $8.40
1 avocado: $8.40 ÷ 4 = $2.10
$2.10 − $0.75 = $1.35

An avocado costs $1.35 more than a zucchini.

Workbook

Exercise 15, pp. 65-66

1. (a) 0.4
   (b) 0.3
   (c) 0.3
   (d) 0.4
   (e) 0.4
   (f) 0.6
   (g) $0.70
   (h) $0.60

2. (a) 0.06
   (b) 0.05
   (c) 0.04
   (d) 0.06
   (e) 0.06
   (f) 0.06
   (g) $0.09
   (h) $0.05

Exercise 16, p. 67

1. (a) 0.24   (b) 0.21
   (c) 0.13   (d) 0.19
   (e) 0.28   (f) 0.19
   (g) 0.13   (h) 0.12

Exercise 17, pp. 68-70

1. (a) 4.13   (b) 3.22
   (c) 1.47   (d) 2.68
   (e) 22.75   (f) 5.27
   (g) 20.14   (h) 7.05

2. (a) $1.05   (b) $1.15
   (c) $1.45   (d) $1.35
   (e) $1.15   (f) $1.09
   (g) $2.55   (h) $1.75

3. (a) 4.85   (b) 15.15
   (c) 11.75   (d) 9.72
   (e) 37.5   (f) 3.25
   (g) 0.25   (h) 29.35

Exercise 18, 71-72

1. (a) 1.4   (b) 0.75
   (c) 0.25   (d) 0.95
   (e) 1.24   (f) 1.25
   (g) 8.25   (h) 5.85

2. (a) $0.95
   (b) $0.85
   (c) $0.35
   (d) $1.05

Exercise 19, p. 73

1. $35 \div 7 = 5$ $60 \div 3 = 20$ $21 \div 3 = 7$ $30 \div 5 = 6$
   4.6   20.3   7.6   6.0

   $30 \div 6 = 5$ $27 \div 9 = 3$ $36 \div 4 = 9$ $16 \div 8 = 2$
   5.5   3.2   9.3   2.2

   9

Exercise 20, pp. 74-76

1. 4 units = 1.48 m
   1 unit = 1.48 m $\div$ 4 = **0.37 m**
   Each piece is 0.37 m long.

2. 3 kg shrimp: $20.40
   1 kg shrimp: $20.40 $\div$ 3 = **$6.80**
   1 kg of shrimp costs $6.80.

3. 5 units = $28.25
   1 unit = = $28.25 $\div$ 5 = **$5.65**
   Holly spent $5.65.

4. ($3.15 + $4.65) $\div$ 2 =$7.80 $\div$ 2 = **$3.90**
   Each girl paid $3.90.

5.
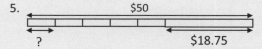

   ($50 − $18.75) $\div$ 5 = $31.25 $\div$ 5 = **$6.25**
   1 kg of grapes cost $6.25.

6. (2.7 lb − 1.2 lb) $\div$ 5 = 1.5 lb $\div$ 5 = **0.3 lb**
   Each block of butter weighs 0.3 lb.

7. Total paint:10.5 ℓ + 15.5 ℓ = 26 ℓ
   Amount of paint in each can: 26 ℓ $\div$ 4 = **6.5 ℓ**
   Each can has 6.5 ℓ of paint.

Review 7

Review

Review 7, pp. 70-73

Workbook

Review 7, pp. 77-82 (answers p. 67)

Tests

Tests, Units 1-7 Cumulative Tests A and B, pp. 37-44

Enrichment

See if your student can solve the following problem.

⇒ 5 avocados and 4 zucchini together cost $18.00. 4 avocados and 5 zucchini together cost $17.10. How much more does each avocado cost than each zucchini? (All avocados cost the same and all zucchini cost the same.)

Start with a diagram:

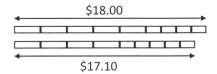

$18.00

$17.10

Let your student work this out on her own. If she did the previous enrichment problems, she might immediately try to make equal numbers of avocados. This involves multiplying the first bar by 4 and the second by 5, so there are 20 avocados in both. After a lot of calculations she will find that the cost of 1 zucchini is $1.50 and the cost of 1 avocado is $2.40. However, it is much easier to solve this problem by rearranging the units. Then, we will find that the difference in the two bars is the same as the difference in one avocado and one zucchini, so only one calculation is needed to find the answer:

1. (a) 124.66
   (b) 124.57
   (c) 124.46
   (d) 124.55

2. (a) >   (b) <
   (c) <   (d) =

3. (a) 100   (b) 108
   (c) 10   (d) 1

4. (a) $35   (b) 9 kg   (c) 10 m

5. (a) 62.3   (b) 19.95   (c) 32.97

6. (a) 8000   (b) 80   (c) 8   (d) 80,000

7. −30, −6, 2, 20

8. 36

9. radius
   center
   10 cm

10. (a) $35 - 15 \div 5 = n - 3$
        $\underline{35} - 3 = n - 3$
        $n = 35$
    (b) $123 \div 2 = 61 + n$
        $61.\underline{5} = 61 + n$
        $n = 0.5$

    (c) $50 = (2 \times n) \times 2$
        $25 \times 2 = (2 \times n) \times 2$
        $2 \times (\underline{12.5}) \times 2 = (2 \times n) \times 2$
        $n = 12.5$
    (d) $n \times 100 = 23{,}800$
        $n = 238$

    (e) $36 \times 25 = n \times 4 \times 25$
        $\underline{6} \times 4 \times 25 = n \times 4 \times 25$
        $n = 6$
    (f) $\frac{5}{8} + \frac{n}{16} = 1$
        $\frac{10}{16} + \frac{n}{16} = 1$
        $n = 6$

11. $8 \times ? = 456$
    $456 \div 8 = \mathbf{57}$
    The other number is 57.

12. (a) $\frac{4}{12} + \frac{9}{12} = 1\frac{1}{12}$
    (b) $\frac{3}{9} + \frac{5}{9} = \frac{8}{9}$

    (c) $\frac{5}{6} + \frac{3}{6} = 1\frac{2}{6} = 1\frac{1}{3}$
    (d) $\frac{8}{10} + \frac{7}{10} = 1\frac{5}{10} = 1\frac{1}{2}$

13. (a) $\frac{11}{12} - \frac{3}{12} = \frac{8}{12} = \frac{2}{3}$
    (b) $2\frac{4}{7}$

    (c) $\frac{4}{8} - \frac{1}{8} = \frac{3}{8}$
    (d) $5\frac{5}{6}$

(continued next page)

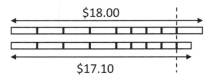

$18.00 − $17.10 = $0.90

An avocado costs $0.90 more than a zucchini.

After showing your student this solution, if you need to, ask her if she can find an easier way to solve the previous problem:

⇒ 2 avocados and 5 zucchini together cost $7.95. 4 avocados and 3 zucchini together cost $10.65. How much more does each avocado cost than each zucchini? (All avocados cost the same and all zucchini cost the same).

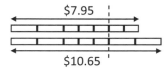

The difference between the two bars, when the units have been rearranged, is the difference between 2 avocados and 2 zucchini. The difference between one avocado and one zucchini is half of that.

$10.65 − $7.95 = $2.70
$2.70 ÷ 2 = $1.35

The purpose of these enrichment problems is to show that it is better to study the problem and solve it using logic and reason rather than just use a set of steps that worked for a similar problem. An approach based on the information in the problem can give an easier and quicker solution. The more practice your student gets in solving different types of problems without being given a set of specific steps, or looking at a solution someone else came up with in a some solution manual, the more he will develop the problem solving skills needed to solve any kind of math reasoning problem .

14. (a) $\frac{1}{12}, \frac{1}{3}, \frac{5}{6}$ (b) $1\frac{1}{4}, 1\frac{3}{4}, \frac{9}{4}$

    (c) $1\frac{3}{5}, 3, \frac{9}{2}$ (d) $2\frac{1}{5}, \frac{9}{4}, \frac{20}{6}$

15. trapezoid

16. D

17. $36,000

18. $77

19. (a) 6.7   (b) 14.3   (c) 4.9

20. Width = 50 cm^2 ÷ 10 cm = **5 cm**

21. Length of one side = 36 in. ÷ 4 = **9 in.**

22. 34 yd ÷ 8 = **4.25 yd**

    Each piece is 4.25 yd long. (or $4\frac{1}{4}$ yd)

23. $0.65 x 6 = **$3.90**
    He paid $3.90.

24. 4.8 m ÷ 8 = **0.6 m**.
    Each piece is 0.6 m long.

25. $\frac{7}{10}$ of the pole was not painted.

26. $\frac{3}{5}$ km x 4 = $\frac{3}{5}$ km + $\frac{3}{5}$ km + $\frac{3}{5}$ km + $\frac{3}{5}$ km

           = $\frac{3\times4}{5}$ km = $\frac{12}{5}$ km = **$2\frac{2}{5}$ km**

    The perimeter of the garden is $2\frac{2}{5}$ km.

27. $\frac{3}{8} = \frac{6}{16}$; **6 out of 16 slices.**

    Or: $\frac{3}{8}$ x 16 = 6

    She gave 6 slices away.

28. $\frac{3}{8}$ were girls.   200

    8 units = 200

    1 unit = $\frac{200}{8}$

    3 units = $\frac{200}{8}$ x 3 = 25 x 3 = **75**

    There were 75 girls.

29. $\frac{3}{4}$ of 6 ℓ = 3 x $\frac{6}{4}$ ℓ = 3 x 1.5 ℓ = **4.5 ℓ**
               (or change liters to milliliters)
    The bucket contains 4.5 ℓ of water.

30. $35.90 + $28.50 = $64.40
    $64.40 − $58.70 = **$5.70**
    He needs $5.70 more.

Workbook

Review 7, pp. 77-82

1. 98,510

2. 100; 1000; 10,000; 100,000; 1,000,000

3. $\frac{6}{10}$ or 0.6

4. 9

5. (a) 48,230
   (b) 70.54

6. AD

7. $4

8. $\frac{1100}{2000} = \frac{11}{20}$

9. (a) 5.25
   (b) 16.8

10. (a) $\frac{17}{20}$
    (b) $2\frac{2}{5}$

11. 5

12. (a) 495   (b) 30

13. 6.3

14. $35

15. (a) 1.3   (b) 1
    (c) 0.1   (d) 0.3
    (e) 0.92   (f) 1.6
    (g) 0.08

16. (a) 17.7
    (b) 76.8

17. (a) 6.05
    (b) 3.7
    (c) 0.61
    (d) 6.7

18. (a) >   (b) =
    (c) >

19. (a) $12 - (3 \times 2) + 9 = 15$ Or
        $12 - 3 \times 2 + 9 = 15$ (no parentheses needed)
    (b) $(12 - 3) \times (2 + 9) = 99$

20. (a) 113.3
    (b) 3.16
    (c) 6.21
    (d) 239.26
    (e) 3.69

21. 3 apples: $1.56
    6 apples are about $3,
    so Alice can buy **9** apples with $5.
    Or, 1 apple: $1.56 ÷ 3 = $0.52
    $5 is about 10 times 1 apple.
    10 apples = $5.20
    Alice can buy 9 apples with $5.

22. 10:15 a.m.

23. 4 km 360 m − 1 km 250 m = **3 km 110 m**

24. Total buns = 98 + 42 = 140; $\frac{42}{140} = \frac{3}{10}$

25. $\frac{2}{5}$

26. 2 lb − $\frac{1}{4}$ lb = **1$\frac{3}{4}$ lb**

27. $\frac{3}{8}$ of $24 = **$9**

28. 2 x (8.5 cm + 4.8 cm) = 2 x (13.3 cm) = **26.6 cm**

29. Width = 1 unit
    Perimeter = 6 units
    6 units = 95.4 m
    1 unit = 95.4 m ÷ 6 = 15.9 m
    2 units = 15.9 m x 2 = **31.8 m**

30. Perimeter: 2 x (35 m + 24 m) = 2 x 59 m = 118 m
    Cost of fencing: 118 x $10 = **$1180**

31. Width: 35 yd^2 ÷ 7 yd = **5 yd**

32. 134°

33. (a) CD // IJ   (b) GH ⊥ PQ

34. Width = 1 unit
    Perimeter = 6 units = 48 in.
    Length = 2 units = 48 in. ÷ 3 = **16 in.**
    The length of the rectangle is 16 in.

35. 1 unit = boys
    6 units = 84
    1 unit = 84 ÷ 6 = **14**
    There are 14 boys.

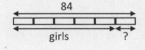

Unit 8 – Congruent and Symmetric Figures

Chapter 1 – Congruent Figures

Objectives

♦ Identify congruent figures.
♦ Name corresponding vertices, sides, and angles in congruent figures.

Materials

♦ Cut-out shapes: Two congruent polygons.
♦ Appendix pp. a11-a12

Vocabulary

♦ Congruent
♦ Corresponding sides
♦ Corresponding vertices
♦ Corresponding angles

Notes

In *Primary Mathematics* 4A, students reviewed the names and properties of various common plane figures and identified them based on their sides and angles.

In this chapter, your student will identify congruent figures and corresponding sides and angles of two congruent figures. This chapter will be restricted to closed plane figures only.

Congruent figures have the same shape and size. If two figures are congruent, then the vertices, sides, and angles that match are called **corresponding vertices**, **corresponding sides** and **corresponding angles**. In the figure below, triangle ABC is congruent to triangle DFE. Angle A corresponds to angle D, angle B corresponds to angle F, and angle C corresponds to angle E. Side AB corresponds to side DF, side BC corresponds to side FE, and side CA corresponds to side ED. Triangle DFE has been both flipped and rotated compared to triangle ABC.

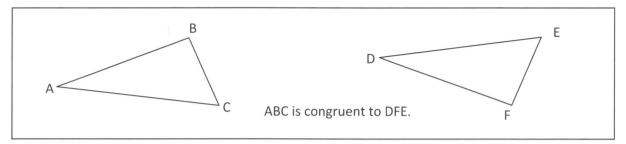

ABC is congruent to DFE.

Congruent figures are named in the order of their corresponding parts. For these two triangles, we say "triangle ABC is congruent to triangle DFE," not "triangle ABC is congruent to triangle DEF."

One way to determine if two figures are congruent is to trace around one of the triangles and place it over the other. If the angles and sides match exactly, then the two are congruent. Another way is to measure the corresponding parts. If the corresponding angles and sides are the same, then the figures are congruent. In *Primary Mathematics*, the figures are placed on a square grid of evenly spaced dots so that the student can use the dots to measure the sides and determine if the sides and angles are the same, rather than having to use a ruler and protractor.

(1) Identify congruent figures

Activity

Before the lesson, cut out two congruent 4-sided or 5-sided figures. You can fold an index card in half and cut both out at the same time. Show the two figures to your student and ask her to compare them. She might say they are the same. Tell her that that figures that have the same size and shape are called *congruent* figures. Set one down and manipulate the other various ways, such as flipping and rotating it, and ask if they are still the same.

Lay one of the figures over the other and label the vertices on both sides, using different letters for each figure. If labeling them with letters in alphabetic order, go clockwise around one and counter-clockwise around the other so your student does not use alphabetical order to find the answers. Then flip and rotate one figure with respect to the other. Tell your student that angles or sides that are the same for two congruent figures are called *corresponding* angles or sides. Name a vertex, angle, or side on one of the figures and ask him to name the corresponding angle or side on the other figure. When naming corresponding sides, the order should correspond. In the example on the right, DE corresponds to ON, not NO. If he has trouble, start by rotating the figure without flipping it.

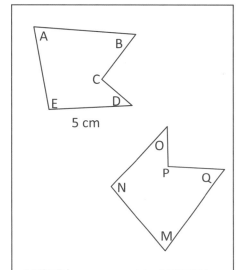

ABCDE is congruent to MQPON

∠C corresponds to ∠P.
DE corresponds to ON.
If DE is 5 cm, then ON is 5 cm.
If ∠A is 85°, then ∠M is 85°.

List the vertices of one figure going clockwise around the figure. Get your student to name the other figure using the same order for corresponding vertices.

You can repeat with more complex figures or ones with more sides. You can write a length or angle size on one figure and ask which length or angle is the same on the other.

Discussion

Concept p. 74

Have your student match the figures that are congruent. You can make 2 copies of appendix p. a11 and cut out the shapes from one copy to fit on top of the shapes of the other copy.

Congruent pairs:
A and J   B and F
G and K
C and E
H and I

Practice

Tasks 1-2, p. 75-76

Workbook

Exercise 1, p. 83-85 (answers p. 75)

1. (b), (c), and (e)
2. (a) A → H   B → E
       C → F   D → G
   (b) DA
   (c) AB and HE
       BC and EF
       CD and FG

Reinforcement

Extra Practice, Unit 8, Exercise 1, pp. 135-136

Label some of the vertices on the figures on appendix p. a11. Tell your student a vertex or a side and ask him to name the corresponding side on the congruent figure. C and E would be good ones to do this with. You can also write down the vertices of one figure, and ask him to write down those of the congruent figure in the corresponding order.

Use a copy of appendix p. a12. Draw a figure on it using the grid, and ask your student to draw a congruent figure with a different orientation.

Test

Tests, Unit 8, 1A and 1B, pp. 45-51

Chapter 2 – Tiling Patterns

Objectives

- Identify the congruent shapes used in a tessellation.
- Continue a tessellation on dot paper.
- Determine whether a shape can tessellate or not.
- Make different tessellations with a given shape.

Material

- Square and isometric dot paper (appendix pp. a12-a13)
- Appendix pp. a14-a16

Vocabulary

- Tiling pattern
- Tessellation

Notes

In this chapter, your student will investigate a practical aspect of the use of congruent shapes and gain more experience with manipulating polygons physically and mentally to determine if they are congruent.

A **tessellation** is an arrangement of congruent shapes on a flat surface. They are geometric patterns that are made of one or more shapes that are fitted together to make a repeating pattern. This ancient form of decoration dates back to the 4th century B.C. A pattern is a tessellation if it is made of one or more shapes that can be extended in every direction to cover a surface and the pattern pieces fit together without any gaps or overlapping.

Your student will need to determine if a given shape can tessellate. It is not necessary for her to come up with a generalization about which shapes can tessellate or not.

For your information, there are only 3 regular polygons (polygons with equal sides and angles) that will always form a tessellation using only one shape: a square, an equilateral triangle, and a regular hexagon. This is because their angles are a factor of 360°, the sum of angles at a point. The angles of a hexagon are 120°, so three hexagons can make a complete rotation.

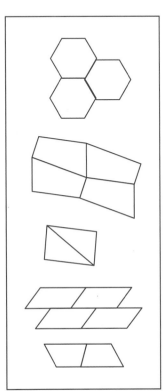

A general quadrilateral will tessellate by having all 4 different corners meet at a point, since the sum of the angles of a quadrilateral is 360°, and the sides will match up in length.

All triangles will tessellate. A congruent triangle can be rotated so that corresponding sides fit against each other, forming a quadrilateral.

Rectangles will tessellate in a number of different ways. A parallelogram will tessellate by translation. Trapezoids will tessellate because two of them make a parallelogram.

Many other figures can tessellate. The enrichment activity for this unit shows how some figures which tessellate can be constructed from other figures which are known to tessellate.

(1) Identify and extend tessellations

Discussion

Concept p. 77
Task 1, p. 78

Your student should notice that all the shapes in these tessellations are congruent, i.e. the shape repeats, that there are no gaps in the design, and that the shape can be added to extend the design in any direction. Tell him that a design formed by fitting shapes together to make a repeating pattern with no gaps or overlaps is called a *tessellation*.

Have your student look around the environment for examples of tessellations, such as tiled floors or designs on fabric. Many of these will consist of more than one shape that is repeated in the pattern. Tell her that in the textbook and workbook, there will only be tessellations of one shape.

Practice

Draw a shape that can tessellate on the square-dot paper and have your student create a tessellation by drawing congruent shapes. You can use any quadrilateral or triangle. Repeat with a more complicated shape such as some on appendix p. a14. If your student has trouble, trace the shape and cut it out so that he can move it around and flip it over to see how it would fit against the shape drawn on paper.

Workbook

Exercise 2, pp. 86-88 (answers p. 75)

Enrichment

Have your student explore various tessellations on the internet. The artwork of M.C. Escher consists of many interesting tessellations.

If your student likes being artistic, have her create some tessellations, possibly on the computer.

(2) Determine if a shape can tessellate

Discussion

Task 2, p. 79

This task first shows two types of shapes, one that can tessellate and one that cannot. Then your student is given 4 shapes and asked to find out which can tessellate. Rather than making 12 copies of each, he could make just one or a few copies of the shape on square dot paper. Appendix p. a16 has a copy of each of the shapes on this page of the textbook, enlarged to fit on copies of the square dot paper on appendix p. a12.

Practice

Give your student some other shapes to see if they tessellate. You can use the shapes from appendix pp. a14-a15.

Workbook

Exercise 3, p. 89-90 (answers p. 75)

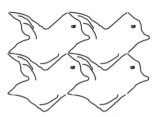

2. A, B, and D can tessellate. C cannot.

A:

B:

C:

D:

Enrichment

M.C. Escher is an artist that has created a wide variety of shapes that can tessellate. Many of his creations are things in nature, like fish, birds, insects, or dinosaurs. If your student is interested, you can show her the following methods for creating Escher types of shapes. Once the tessellating shape is created, she can draw details in it to make it look like an animal or person or other object, then copy enough to fill up a sheet of paper, or trace the shape on paper.

You can modify a rectangle by cutting one side and sliding it over to the other side.

You can also modify the other side in the same way.

Tessellations can also be created from triangles by marking the midpoint on each side, cutting a piece from one half of one side, and rotating it 180° around the midpoint. You can do this with each side.

 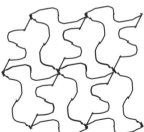

(3) Make different tessellations from the same shape

Discussion

Task 3, p. 80

This task first shows that different tessellations can be made from one kind of shape. You can have your student draw additional possible tessellations of the rectangle using the square dot paper and rectangles whose area is two square units, or copied shapes of rectangles. Have him do some other pattern than simply even rows and columns.

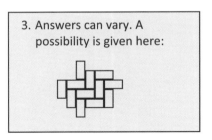

3. Answers can vary. A possibility is given here:

Practice

Task 4, p. 80

Appendix p. a16 has a copy of each of the shapes on this page of the textbook, enlarged to fit on copies of the square dot paper on appendix p. a12.

Workbook

Exercise 4, p. 91-94 (answers p. 75)

Reinforcement

Extra Practice, Unit 8, Exercise 2, pp. 137-140

Test

Tests, Unit 8, 2A and 2B, pp. 53-56

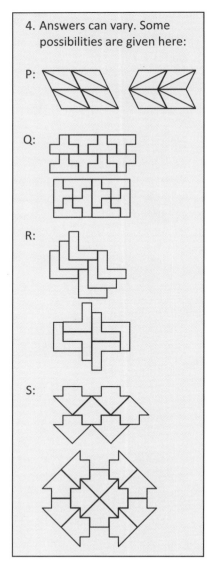

4. Answers can vary. Some possibilities are given here:

P:

Q:

R:

S:

Workbook

Exercise 1, pp. 83-85

1. A and G
   B and C
   E and D

2. Answers can vary. Check answers.

3. Assume the figures are not symmetric, and are just rotated relative to each other.
   (a) F
   (b) GH (error in first printing, should say side BC)
   (c) CD
   (d) B
   (e) IJ

Exercise 2, pp. 86-88

1. Check that your student has colored in a single shape in each.

2. (a)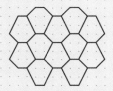

   (b)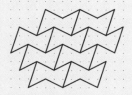

   (c)

   (d)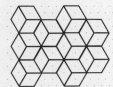

Exercise 3, pp. 89-90

1. (a) No   (b) Yes

  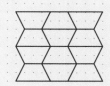

   (c) Yes   (d) No

  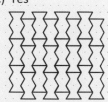

Exercise 4, pp. 91-94

1. Answers can vary
   (a)   (b)

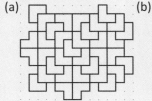 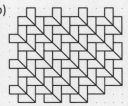

   (c)   (d)

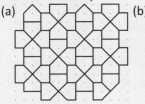 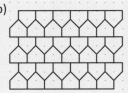

2. Answers can vary.
   (a)   (b)

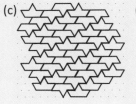 

   (c)   (d)

Chapter 3 – Line Symmetry

Objectives

- Identify symmetric figures with line symmetry.
- Draw or make cut-outs of symmetric figures.
- Recognize that two halves of a symmetric figure are congruent.
- Identify and draw lines of symmetry in a figure.
- Investigate properties of symmetry in common geometric shapes.
- Complete a symmetric figure, given one half of the figure and the line of symmetry.

Material

- Small mirrors
- Appendix pp. a17-a20

Vocabulary

- Symmetric
- Reflection
- Line of symmetry

Notes

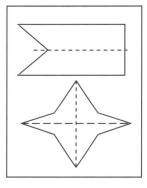

In this chapter, your student will learn to identify figures with line symmetry.

Plane figures that have a line of symmetry are called **symmetric** figures. The **line of symmetry** divides the figure into two parts, each of which is a **reflection** or mirror image of the other. When one part is flipped about the line of symmetry, it matches the other part. The two figures at the right are symmetric figures, and the dashed lines are lines of symmetry. Some figures, like the second one, have more than one line of symmetry. Your student will not be required to find the total number of lines of symmetry in a figure at this level, but she should recognize that there may be more than one line of symmetry.

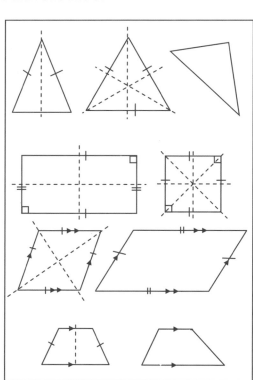

Your student will be investigating lines of symmetry in triangles and some quadrilaterals in this chapter. He should know which sides are equal or parallel in these figures, and the meaning of the notches (equal sides) and arrows (parallel sides) on the figures and the little square in the corner (right angle), which he learned in *Primary Mathematics* 4A. Other angle properties of these figures will be covered in *Primary Mathematics* 5.

For triangles, an isosceles triangle has one line of symmetry, an equilateral triangle has 3 lines of symmetry, and a scalene triangle has no lines of symmetry.

For parallelograms, a rectangle and a rhombus have two lines of symmetry and a square has 4 lines of symmetry; all other parallelograms have no lines of symmetry.

If a trapezoid is equilateral (the two non-parallel sides are equal) then it has one line of symmetry, otherwise it has no lines of symmetry.

(1) Identify symmetric figures

Discussion

Concept p. 81

Ask your student what she can tell you about the two halves of the figures on this page. Tell her each half is a mirror image or *reflection* of the other. If you have a small mirror with a straight side, you can line up the mirror with the line of symmetry in each of the figures and have her see that the reflected image in the mirror is the same as the other half of the figure behind the mirror.

There are many examples of symmetry in the environment, but often examples in the environment are three dimensional. You can tell your student that solid objects are symmetric if you could cut them in half in such a way that one half is the "mirror image" of the other. Such shapes have "bilateral symmetry." With natural objects, such as animals, the two halves are not exactly identical, but they still are said to have bilateral symmetry if the two halves have the same parts. A person has bilateral symmetry.

Tasks 1-2, p. 82

Get your student to cut out some symmetric figures. The main concept he needs to understand is that the two parts of a figure are reflections (mirror images) of each other and that if you fold a figure on the line of symmetry, one half fits over the other exactly so that all the corresponding angles and sides are equal.

Activity

Have your student cut the figures from small sheets of paper or index cards folded in half, or you can create some complex symmetric figures with minor differences in advance. Keep the page from which the figures were cut folded, but open the cut-out figure. Mix them up and change the orientation and ask her to match each cut-out with its source without moving it.

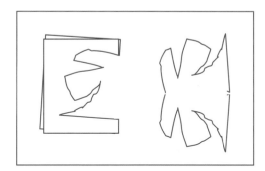

Workbook

Exercise 5, pp. 95-96 (answers p. 83)

(2) Identify lines of symmetry in geometric shapes

Discussion

Tasks 3-7, pp. 83-85

Provide your student with paper cut-out rectangles, isosceles, equilateral, and scalene triangles, parallelograms, a rhombus, and two kinds of trapezoids to fold while you discuss these tasks. You can copy and cut out the figures on appendix p. a17. After doing the tasks associated with each type of shape, you can have him also find out how many lines of symmetry each figure has. After doing Task 7, you can give him a paper square and ask him to find out how many lines of symmetry it has (4).

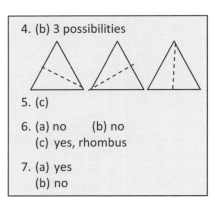

4. (b) 3 possibilities

5. (c)

6. (a) no   (b) no
   (c) yes, rhombus

7. (a) yes
   (b) no

3: This task shows that a figure can have more than one line of symmetry, and that just because a line divides a figure into two congruent halves does not mean that it is a line of symmetry. In the last figure on this page, the diagonal of the rectangle is not a line of symmetry even though the two halves are congruent.

4: Have your student fold the triangle cut-outs to find lines of symmetry. Point out that there are three lines of symmetry for an equilateral triangle.

5: Ask your student why the triangle in (c) has a line of symmetry, but the ones in (a) and (b) do not. The one in (c) is an isosceles triangle, as well as a right triangle.

6: Again, have your student fold some paper cutouts of parallelograms similar to the one in this task. Although she can often determine if a given line is a line of symmetry by inspection, the one on the parallelogram in (a) may seem to be a line of symmetry since a similar line from midpoints of the sides in a rectangle is a line of symmetry. The lines on the two parallelograms in (a) and (b) do divide the shapes into congruent halves, but they are not mirror images of each other.

Practice

Task 8, p. 86

Your student should be able to answer these by inspection. If not, have him trace them and cut them out.

8. (a) yes   (b) yes
   (c) no   (d) yes

Workbook

Exercise 6, pp. 97-98 (answers p. 83)

Reinforcement

Have your student draw some other symmetric plane figures using graph paper, square dot paper, or isometric dot paper.

Enrichment

Your student may be able to find lines of symmetry by simple inspection, or by tracing the figure, cutting it out, and folding along the proposed line of symmetry. However, since she has learned in *Primary Mathematics* 4A how to draw perpendicular lines using a set-square and ruler, or on square grid paper, you can show her how to determine if a given line is a line of symmetry by having her draw perpendicular lines to the line of symmetry.

If a given line is a line of symmetry, then a line drawn from a point on one side to its corresponding point on the other side is perpendicular to the line of symmetry and is bisected (cut in half) by the line of symmetry.

So if we can draw a line perpendicular to the proposed line of symmetry through a specific point on one side, such as a vertex, and this line does not go through the corresponding point on the other side, such as the opposite vertex, then the proposed line of symmetry is not a line of symmetry.

For example, in the parallelogram at the right, the sides on both sides of the dashed line look the same, but the dashed line is not a line of symmetry; a line perpendicular to the dashed line which goes through one corner does not go through the opposite corner.

If you have a transparent ruler, it can easily be placed perpendicular to the proposed lines of symmetry in the textbook by lining them up with one of the marks on the ruler, which are perpendicular to the sides of the ruler. If you do not have a transparent ruler, you can have your student use a set-square to put the ruler perpendicular to the proposed line of symmetry.

(3) Complete a symmetric figure

Discussion

Task 9, p. 86

Provide your student with graph paper and have him copy the figure and then complete it using AB as the line of symmetry.

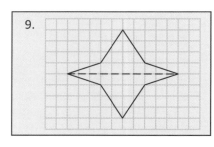

Practice

Have your student complete the figures on appendix p. a19 using the dotted lines for lines of symmetry.

Workbook

Exercise 7, p. 99-100 (answers p. 83)

Reinforcement

Extra Practice, Unit 8, Exercise 3, pp. 141-144

Test

Tests, Unit 8, 3A and 3B, pp. 57-63

Enrichment

If your student has not already discovered this on her own, show her that she can find where a corner or vertex of the figure should be by first drawing or imagining a perpendicular line from a vertex of the figure on one side to the line of symmetry, and then extending it the same distance on the other side. Use the figure from Task 9 to show how this is possible, and then guide her in using this approach with the first figure on appendix p. a20, as shown at the right. Then let her complete the rest of the figures on that page.

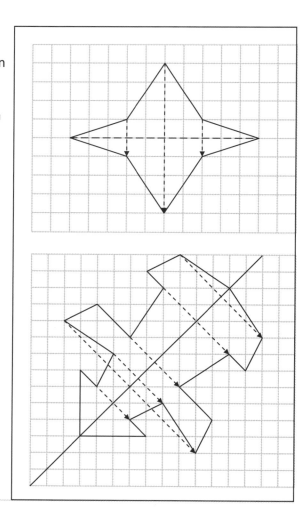

Chapter 4 – Rotational Symmetry

Objectives

♦ Identify figures with rotational symmetry.

Material

♦ Appendix pp. a21 - a22
♦ Cardboard, tack or pin

Vocabulary

♦ Rotational symmetry

Notes

In this chapter, your student will be introduced to rotational symmetry for plane figures.

A figure has **rotational symmetry** if there is a point in the figure that it can be turned around a certain number of degrees less than 360° (one full turn) and still look exactly the same. The point it is rotated around is called the center of rotation. The figure at the right has rotational symmetry because a tracing of it can be rotated 120° twice and each time will match the original figure.

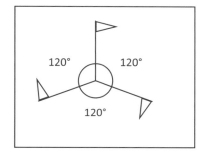

For your information, the number of positions a figure can be rotated to, without bringing in any changes to the way it looks originally, is called its order of rotational symmetry. If a figure has a rotational symmetry of 2, that means there are 2 positions it can be in and look exactly the same as it does in the original position. Any figure can be rotated one full turn to its original figure. In order to have rotational symmetry, its order of rotational symmetry must be at least 2. The figure at the right has an order of rotational symmetry of 3.

At this level, your student will only determine whether a figure has rotational symmetry. She will not be required to find the order of rotational symmetry.

(1) Identify figures with rotational symmetry

Discussion

Concept p. 87

To do the activity, you can make two copies of the figure on appendix p. a21. (It is slightly different than the figure in the textbook so that cutting it out will not separate the two halves. Alternately, you can make one copy onto transparent paper.) Cut one out and lay it on top of the other. Put a pin or tack through both figures at the center and into some cardboard.

> A will lie on top of D when the rotated figure matches the original figure.

There are many examples of rotational symmetry your student can look for if we ignore the color and just focus on the shape such as dartboards, hub caps, Ferris wheels, or playing cards. Your student can also look for rotational symmetry in nature, which will not be exact but still is obviously based on rotational symmetry, such as flowers, starfish, and the cross section of some fruit.

Have your student look at the figures on appendix p. 22 and determine whether they have line symmetry, rotational symmetry, or no symmetry. She should take into account shading when determining the symmetry of the figures.

> Appendix p. a22
>
> (a) line, rotational   (b) rotational
> (c) line, rotational   (d) line, rotational
> (e) rotational   (f) rotational
> (g) none   (h) line
> (i) line, rotational   (j) line, rotational
> (k) line   (l) none

Practice

Tasks 1-4, p. 88

> 1. 1st, 3rd, and 4th figures
> 2. FG
> 3. parallelogram, rhombus, rectangle, and square
> 4. 3rd triangle (equilateral triangle)

Activity

Have your student draw some figures with rotational symmetry, either manually or on the computer. Challenge her to draw a figure with rotational symmetry but no line symmetry.

Fold some paper into fourths or eighths and cut bits out of it and then open it up to see if it has rotational symmetry (make a snowflake).

Workbook

Exercise 8, p. 101 (answers p. 83)

Reinforcement

Extra Practice, Unit 8, Exercise 4, pp. 145-146

Test

Tests, Unit 8, 4A and 4B, pp. 65-68

Workbook

Exercise 5, pp. 95-96

1. (a)

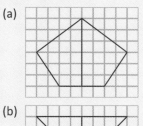

(b)

2.

Exercise 6, pp. 97-98

1. (a)   (b) not symmetrical

(c)   (d)

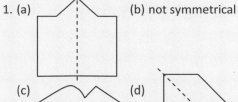

(e)   (f) not symmetrical

(g)   (h)

2. (a) yes   (b) no
   (c) yes   (d) yes
   (e) no   (f) no
   (g) no   (h) yes

Exercise 7, pp. 99-100

1.

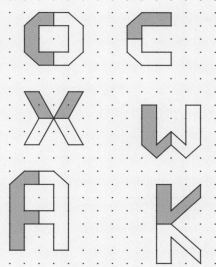

2.

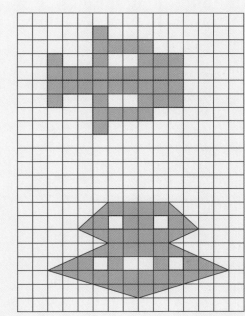

Exercise 8, p. 101

1. Windmill: rotational symmetry
   Butterfly: line symmetry
   Star: line and rotational symmetry
   Bottle: line symmetry
   Hexagon: line and rotational symmetry

2. Check drawings.

Review 8

Review

Review 8, pp. 89-92

Workbook

Review 8, pp. 102-106 (answers p. 85)

Tests

Tests, Units 1-8 Cumulative Tests A and B, pp. 69-79

1. (a) 2
   (b) 0.03 or $\frac{3}{100}$

2. (a) 90,504
   (b) 17,541
   (c) −405

3. 2.69

4. (a) 80,300; 82,300
   (b) 5.59, 6.09
   (c) 0, −5

5. (a) 14,680
   (b) 30,083
   (c) 9,900
   (d) 89,301

6. (a) 7.03
   (b) 4.9
   (c) 2.41
   (d) 3.602

7. (a) 14,058; 14,508; 41,058; 41,508
   (b) 0.96, 8.54, 24.3, 72
   (c) −14, −10, −8, 15, 20

8. (a) 0.28   (b) 0.04

9. (a) 40 x (23 − 15)   (b) 38 x 2 − 7 x 4
     40 x 8   76 − 28
     320   **48**

    (c) (36 + 24) ÷ 5   (d) 8 x (43 − 38) ÷ 2
     60 ÷ 5   8 x 5 ÷ 2
     12   40 ÷ 2
                            20

10. (a) <   (8 < 9)
    (b) >   (355 > 335)
    (c) =   (450 = 450)
    (d) >   (5405 > 5045)

11. (a) A: 4490   B: 4540   C: 4620
    (b) P: 2.43   Q: 2.49   R: 2.54

12. 8.5

13. (a) Any 2 of these: 1, 3, 5, 9, 15, 45
    (b) 24, 48, 72...

14. (a) 10.41
    (b) 15,336

15. D does not have a line of symmetry.
    A and D are nets of cubes.

16. (a) line symmetry
    (b) line and rotational symmetry
    (c) rotational symmetry

17. 3560 + 2790 = **6350**
    The larger number is 6350.

18. 1242 in. ÷ 9 = **138 in.**
    The red ribbon is 138 in. long.

19. $1.25 x 6 = **$7.50**
    She spent $7.50.

20. 1.5 qt − 0.75 qt = **0.75 qt**
    0.75 qt more water can be poured into the bottle. (Or 3 cups.)

21. $50 − ($11.90 + $27.35) = $50 − $39.25 = **$10.75**
    He had $10.75 left.

22. $20.35 + $20.35 + $16.85 = **$57.55**
    They saved $57.55 altogether.

23. Paid by students: 18 x $3 = $54
    Paid by Miss Bowles: $72 − $54 = **$18**
    Miss Bowles paid $18.

24.
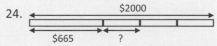

($2000 − $665) ÷ 3 = $1335 ÷ 3 = **$445**
Each microwave oven cost $445.

25. Rent for 2 months: $4500 x 2 = $9000
    $9000 ÷ 4 = **$2250**
    Each man paid $2250.

| | | Workbook | | |

Review 8, pp. 102-106

1. (a) 10,590; 10,050; 9950; 9590; 9190
   (b) 8.3; 7.28; 2.83; 2.05
   (c) 62, 21, −20, −34, −42

2. (a) 57.76
   (b) 4.43
   (c) 20.15
   (d) 282

3. (a) (24 ÷ 6) ÷ 2 + 3 = 5
   or 24 ÷ 6 ÷ 2 + 3 = 5 (parentheses not needed when following order of operations)
   (b) 24 ÷ (6 ÷ 2) + 3 = 11

4. (a) 51.2 (38.59 + 12.62 = 51.21)
   (b) 44 (4.85 x 9 = 43.65)
   (Note: rounding the sum or product is not the same thing as finding an estimate.)

5. $\frac{13}{5}$

6. $4\frac{3}{4}$

7. $4\frac{6}{25}$

8. 6.8

9. 26.08

10. A: 5.78   B: 5.84   C: 5.87

11. (a) 32.82 − 6.45 = **26.37**
    (b) 56.04 − 21.99 = **34.05**
    (c) 17.12 ÷ 8 = **2.14**
    (d) 41.1 x 6 = **246.6**

12. 9 h 35 min

13. 1.7 m − 0.46 m = **1.24 m**

14. (4 + 6) x 48 = 10 x 48 = **480**

15. $\frac{6}{8} = \frac{3}{4}$

16. $\frac{8\text{ hours}}{24\text{ hours}} = \frac{1}{3}$

17. 4 gal − $\frac{3}{5}$ gal = $3\frac{2}{5}$ **gal**

18. $\frac{3}{8}$ of 40 = 3 x $\frac{40}{8}$ = 3 x 5 = **15**

19. (a) 33° (can be calculated: 90° − 57° = 33°)
    (b) 44° (can be calculated: 180° − 136° = 44°)

20. 1 can of peaches costs the same as 2 cans of beans, so 1 can of beans and 2 cans of peaches costs the same as 5 cans of beans.
    $1.20 x 5 = **$6.00**
    The total cost is $6.00.

21.

	Line symmetry	Rotational symmetry
Square	✓	✓
Rectangle	✓	✓
Parallelogram		✓
Rhombus	✓	✓
Trapezoid		
Equilateral triangle	✓	✓
Isosceles triangle	✓	
Scalene triangle		

22. Answers can vary. Two possibilities are shown here.

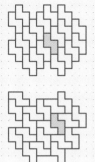

Unit 9 – Coordinate Graphs and Changes in Quantities

Chapter 1 – The Coordinate Grid

Objectives

♦ Identify and graph ordered pairs on a coordinate grid.
♦ Use subtraction to find the length of horizontal and vertical line segments on a coordinate grid.

Materials

♦ Map
♦ Coordinate grids (appendix pp. a23-a24)
♦ Rulers

Vocabulary

♦ Coordinate grid
♦ Axes
♦ *x*-axis
♦ *y*-axis
♦ Origin

♦ Ordered pair
♦ Coordinates
♦ *x*-coordinate
♦ *y*-coordinate
♦ Graph (verb)

Notes

In earlier levels of *Primary Mathematics*, students learned how to locate points on both horizontal and vertical number lines. In this chapter, your student will learn to graph positive points on a plane by using a coordinate grid.

A **coordinate grid** (or plane) consists of a horizontal number line and a vertical number line that intersect at right angles to form a plane. The horizontal line is usually called the ***x*-axis** and the vertical line the ***y*-axis**.

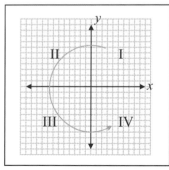

For your information, the coordinate grid is divided into 4 quadrants. In the first quadrant, both coordinates are positive (+,+). The other quadrants are numbered in counterclockwise direction; the second quadrant (−,+), the third quadrant (−,−), and the fourth quadrant (+,−). These quadrants are often designated with Roman numerals. At this level, your student will only be concerned with the first quadrant, and will only graph points where both coordinates are positive. In *Primary Mathematics* 5 and 6, students will graph points in the other quadrants.

We can locate any point on the grid by naming the **coordinates** of the point. These coordinates are **ordered pairs** of numbers, sometimes written within parentheses, such as (2, 4). The first number in the pair, the ***x*-coordinate**, indicates the location on the x-axis. The second number in the pair, the ***y*-coordinate**, indicates the location on the y-axis.

The distance between two points along a horizontal line can be determined by finding the difference between the two *x*-coordinates, and the distance between two points along a vertical line can be determined by finding the difference between the two *y*-coordinates.

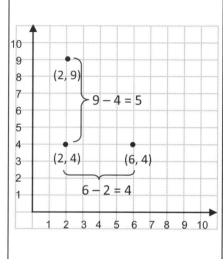

(1) Understand the coordinate grid

Activity

Show your student a map and discuss ways to locate geographical points on the map. Tell her that we can locate something on the map if we are given the *coordinates* of the location. The coordinates tell us the distance east or west and the distance north or south from a particular spot, usually the lower left corner of the map. If you use a globe, you can briefly discuss locating geographical sites using latitude and longitude (which start where the Prime Meridian line crosses the equator.)

Give your student a 10 by 10 coordinate grid with the axes drawn (appendix p. a23). Tell him that the two number lines form a *grid*. With a grid, instead of locating a point along a line, we can locate a point on a flat surface. The number lines are called the *axes* of the grid.

If necessary, review the terms horizontal and vertical and make sure your student knows which number line is horizontal and which one is vertical. (A horizontal line, from the word horizon, is parallel to the horizon, or the bottom edge of the page, and a vertical line is perpendicular to the horizontal line.) Tell her that rather than having to say horizontal or vertical axes each time, the axes are usually labeled with letters, most often with *x* and *y*. By convention, the *x-axis* is the horizontal axis and the *y-axis* is the vertical axis. Label those axes on the grid. Ask your student to identify the point at which these two axes intersect. This point of intersection, which is at 0 on both axes, is called the *origin*.

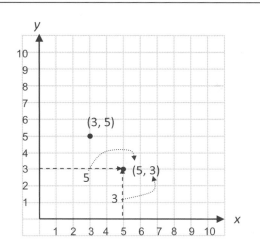

The location of any point is given by (*x, y*), where *x* is the horizontal distance from 0 along the *x*-axis, and *y* is the vertical distance from 0 along the *y*-axis.

Draw a point on the grid at (5, 3). Tell your student we can give the location of the point by how many units it is from the 0 horizontally, that is, along the x-axis, and how many units it is from the 0 vertically, that is, along the y-axis. Have him determine these distances. Write **(5, 3)** next to the point. Tell him that these numbers are called the *coordinates* of the point. The first coordinate is usually called the *x-coordinate* and the second coordinate is usually called the *y-coordinate*. So in general, the location of a point is written as (*x, y*), where *x* is the horizontal distance from 0 and *y* is the vertical distance from 0. Tell him that the order of the numbers is very important. The first number is *always* the distance along the horizontal axis, no matter what this axis is labeled as, so the two numbers together are called an *ordered pair*.

Write the ordered pair **(3, 5)** and guide your student in drawing a point on the grid with those coordinates. Write some other ordered pairs and have her draw a point at the correct location. Emphasize that the first coordinate is the distance along the *x*-axis and the second the distance along the *y*-axis each time. They are easy to mix up initially. It can help to remind her that the order is the same order alphabetically by the labels of the axes, *x* and *y*, but axes are not always labeled and your student still needs to remember which comes first.

Discussion

Concept p. 93

> The monkey's position is (3, 5).

Discuss this page, reviewing the vocabulary. In this example, the banana is at the origin and the units for the grid and the orientation of the grid are arbitrary. The example relates map location to location of a point on a coordinate grid; with maps the orientation is east-west for the horizontal axis, north-south for the vertical axis, and the units are distances. Note that the grid does not really answer the question about who is closer to the banana, for that we would measure the length of a straight line from the banana to the animals.

Tasks 1-2, pp. 94-95

In Task 1 your student needs to label a given point with an ordered pair. In Task 2 he is given the ordered pair and needs to locate the point. In the workbook he will not just be finding which point is at the correct location, but will be graphing ordered pairs. Tell him that when we draw a point for an ordered pair, we say that we *graph* the point. Sometimes the word plot is used, i.e. we plot the points.

1. (a) (5, 4)
   (b) (3, 6)
   (c) (8, 7)
   (d) (2, 2)
   (e) (1, 2)

2. (a) E
   (b) B
   (c) C
   (d) D

Practice

Give your student a coordinate grid. Have her graph the following sets of three points given at the right and connect them in order (i.e. first point to second and second to third). Each set is 3 corners of a parallelogram. For each, have her complete the parallelogram and write down the coordinates for the fourth corner.

(1, 3), (1, 5), (3, 8)	((3, 6))
(4, 3), (3, 1), (7, 1)	((8, 3))
(7, 6), (5, 6), (7, 4)	((9, 4))
(5, 6), (4, 9), (5, 10),	((6,7))

Workbook

Exercise 1, pp. 107-108 (answers p. 91)

Enrichment

Have your student graph some points on a larger grid (appendix p. a24).

Use graph paper and draw axes that include negative numbers and have your student graph or name points in all four quadrants.

If your student likes to draw maps, have him draw one on a coordinate grid. Then, using a blank grid, have him give you verbal directions for drawing the same map, such as, "put a mountain at (4, 5)" or "have a mountain range go from (1, 9) to (3, 8)." See how well his directions allowed you to reproduce his map.

(2) Find the distance between horizontal or vertical points

Activity

Draw a horizontal number line. Tell your student that the distance is marked in units. Mark two points on the line and ask her for the distance between the points. Then ask her how she found the distance. She could simply count the units, but she could also subtract the smaller number from the larger number. If the distances were larger, say the two points were 21 and 89, we would use subtraction to find the distance between them.

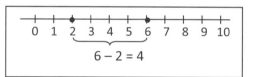

Use a coordinate grid and draw two points along a horizontal line on the grid and label them. Ask your student how we would find the distance between the two points. Again, in this case we can simply count the units. We can also subtract the x-coordinates of the two points. Have your student examine the two ordered pairs and tell you what he notices. The x-coordinates are different but the y-coordinates are the same. If the y-coordinates are the same, then we know a line between the two points is horizontal, since both are the same distance up. If they are, then we can find the distance between the two points by subtracting the x-coordinates, which tells us where the two points are along the x-axis.

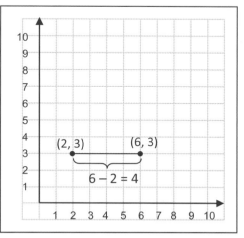

Repeat with two points along a vertical line. This time, the two x-coordinates are the same, which tells us that a line between them is vertical. We can then subtract the two y-coordinates in order to determine how far apart the two points are.

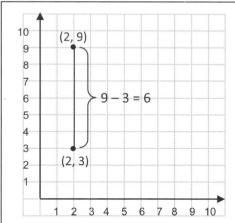

Point out that we can only find the distance between two points this way if either both the x-coordinates or both the y-coordinates are the same. Graph the points (1, 8) and (8, 1). We cannot find the distance between them using simple subtraction since neither the x-coordinates nor the y-coordinates are the same, so a line drawn between them is not parallel to either axis. If necessary, have your student measure the line to see that there is no way to calculate the distance between the points using simple subtraction of the coordinates. (Students will learn how to find the distance between any two points at the secondary level.)

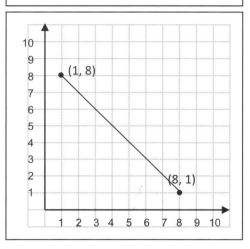

Discussion

Task 3, p. 95

Ask your student to explain her answer to (c). Then ask her to tell you which coordinates we could subtract to find the distance between Bill and Manuel's houses, and why. Ask her which coordinates are the same if two points are vertical to each other, and which if the two points are horizontal to each other.

Practice

Tasks 4-6, p. 96

5, 6: Provide your student with coordinate grids (appendix p. 24) if needed. He may be able to do these two tasks without graphing the points.

Workbook

Exercise 2, pp. 109-110 (answers p. 91)

Reinforcement

Extra Practice, Unit 9, Exercise 1, pp. 149-150

Write each set of two ordered pairs given at the right and ask your student to tell you if a line between the two points is horizontal, vertical, or neither. If horizontal or vertical, ask for the distance between the two points. You can repeat with larger numbers for some quick subtraction practice.

Test

Tests, Unit 9, 1A and 1B, pp. 81-88

---

3. (a) 5 units
   (b) 6 units
   (c) the second or *y*-coordinates

---

4. (a) (i) 3 units (ii) 4 units
   (b) second
   (c) 14 units

5. (a) 5 units
   (b) the *x*-axis

6. (a) $10 - 3 = 7$
   $6 - 2 = 4$
   $2 \times (7 + 4) = 22$
   22 units

(12, 9) (12, 30)   vertical, 21

(29, 15) (15, 20)   neither

(19, 7) (9, 7)   horizontal, 10

(24, 12) (34, 24)   neither

Workbook

Exercise 1, pp. 107-108

1. (a) (3, 6)
   (b)

Place	Coordinates
Granite Mountain	(7, 8)
Silver Lake	(1, 7)
Forest	(5, 5)
Power Station	(2, 4)

   (c) Check location marked. (It is above Forest and directly left of Granite Mountain.)
   (d) Check drawing. The river goes from Granite Mountain to one unit to the left and 4 units up.

2.

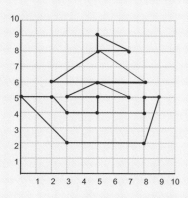

3.

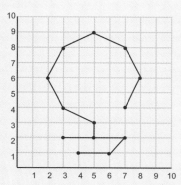

Exercise 2, pp. 109-110

1. (a) (i) 5 units
      (ii) 2 units
      (iii) 9 units
   (b) first
   (c) 6 units (8 − 2 = 6)

2.

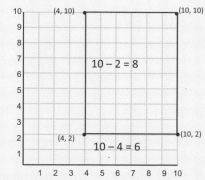

Perimeter = 2 x (6 + 8) = **28 units**

3.

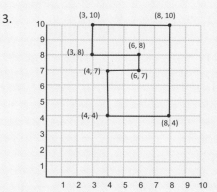

Perimeter = **26 units**

4. (12, 17)

5. (a) (3, 12)
   (b) (9, 12)
   (c) (9, 6)

Chapter 2 – Changes in Quantities

Objectives

♦ Complete a table of values involving two related quantities.
♦ Write a simple linear equation based on a change in quantity.
♦ Use substitution to evaluate an expression with an unknown.

Material

♦ Isometric dot paper (appendix p. a13, optional)
♦ Graph paper (appendix p. a25, optional)

Vocabulary

♦ Equation

Notes

In this chapter, your student will use numerical patterns to relate one quantity to another quantity when the relationship between the two quantities is a direct proportion.

For example, in the following table we see that the second quantity is always two more than the first quantity. Following this pattern, we could complete the table.

Quantity 1	1	2	3	4	5	6
Quantity 2	3	4	5	6	?	?

Quantity 2 is the same as Quantity 1 plus 2.

Your student will then represent the first quantity with a symbol and relate the second quantity to it in a simple equation. For example, for the relationship given in the table above we could represent the first quantity with ■ and the second quantity with ?. Since the second quantity is 2 more than the first, it is also equal to ■ + 2. An equation relating the two quantities is ? = ■ + 2.

Your student will then be able to substitute any value in for ■ and use the equation to get the value for ?.

Students have been writing **equations** representing situations where one number is more than another all along. The only difference here is that one of the numbers is replaced with a symbol since it can change. The answer will be different each time depending on the value of the number represented by ■.

Although relating number patterns to equations with symbols to represent the numbers is one way to introduce students to algebraic reasoning, the *Primary Mathematics* curriculum places a greater emphasis on the approach of drawing bar models and deriving an equation with an unknown, usually called the unit. Thus, more emphasis is placed on translating word problems or "real-life" problems into an equation that can be solved rather than discovering patterns. This chapter and the next are therefore only a brief introduction to functions and graphing functions at a very basic level. These concepts will be covered again each year to algebra, by which time it can be covered in more than just a very basic level.

The concepts in this chapter are relatively abstract, so you may want to allow several days for the lesson.

(1) Write an equation for a change in quantity

Discussion

Concept p. 97

> The last two entries in the table are 7 and 8.
>
> If ▲ is 10, then ■ is 12

Make sure your student understands the geometric pattern shown with the triangles. A congruent copy of the first triangle is flipped and fit against the first triangle. Then an original copy is fit and then a flipped copy so that if the pattern is continued there is a line of triangles.

Your student should be able to fill in the table without needing to see the next two shapes in the pattern, but you can draw them using the isometric dot paper if needed. The purpose of the table is to relate the number of triangles to the number of sides on the outside of the resulting figure. Since the triangle is equilateral and one centimeter is a side of the triangle, this is the same as the perimeter in centimeters.

By looking at the numerical pattern in the table, your student should see that the value for the perimeter is always 2 more than the value for the number of triangles. She may also notice that for each row in the table, the number increases by 1 for both the number of triangles and the perimeter, but the perimeter starts at two more than the number of triangles. This can help her determine what the next two numbers in the table are, but you need to emphasize the relationship between the first number, the number of triangles, and the second, the perimeter, more than the pattern from left to right.

If your student has difficulties with the triangle and square symbols, you can rewrite the equation using a phrase and simplify down to the symbol, as shown at the right. Tell him that once we define what the symbols stand for, we can use them instead of having to write out a phrase describing each quantity.

> Number of triangles + 2 = Number of 1-cm sides
> Triangles + 2 = Sides
> T + 2 = S
> ▲ + 2 = ■
> $x + 2 = y$

Task 1, p. 98

This task essentially uses the pattern in a calendar page to derive an equation for the date a week later than a given date. If your student has trouble, you can use the words "a week from now" and "today" to stand for ◆ and ● initially.

Tell your student that ◆ = ● + 7 is an *equation*. An equation is a mathematical statement where the expression on one side is equal to the expression on the other side. Point out that with ◆ = ● + 7, the value of ◆ depends on what we are given for the value of ●. The value we

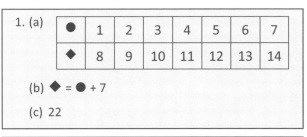

1. (a)

●	1	2	3	4	5	6	7
◆	8	9	10	11	12	13	14

(b) ◆ = ● + 7

(c) 22

> A week from now = Today + 7
> Let ◆ be a week from now and ● be today
> ◆ = ● + 7

need to find, ◆, is now on the left-hand side of the equation. In the previous example, ▲ + 2 = ■, the value we are asked to find is on the right-hand side of the equation. There is no rule for which side to put the value we want to find.

Tell your student that we can use any symbol, as long as we define it. If we say that x stands for today, or the first number in the table, and y stands for a week from now or the number under it in the table, then we can write the equation y = x + 7.

> A week from now = today + 7
> Let y be a week from now and x be today.
> $$y = x + 7$$

Task 2, p. 98

In this task, we are starting with the equation and using it to fill out the values in the table, that is, we are finding the value of A for different values of s. Equations with standard abbreviations, such as A for area, are used a lot for formulas for area and volume of geometric figures.

2. (a)

s	1	2	3	4	5	6
A	1	4	9	16	25	36

(b) 24 x 24 = **576**

This is an interesting pattern and your student will see it again at the secondary level (square numbers). Instead of the number in the second row increasing by 1 as we go along the row, it is increasing much more quickly (exponentially). At first, the area is the same as the side. You can extend this concept by drawing squares on graph paper, such as appendix p. a25, to get a visual idea of how fast the area is increasing compared to the side.

Practice

Tasks 3-4, p. 99

See if your student can do these two tasks independently. Task 4 shows an equation that uses both multiplication and addition. Point out that the equation is not any more complex than ones she has already written with actual numbers rather than symbols. The symbols simply indicate that different values can be used, such as different numbers of shirts. How much she spends (the value of C) depends on how many shirts she buys (the value we use for s).

3. (a) l = **3 x w**
(b) l = 3 x 4 cm = **12 cm**
(c) l = 3 x 8 cm = 24 cm
Perimeter = 2 x (8 cm + 24 cm) = **64 cm**
(d) l = 3 x 10 cm = 30 cm
Area = 10 cm x 30 cm = **300 cm²**

4. (a)

1	2	3	4	5
5	10	**15**	20	25
20	20	20	20	20
25	30	35	40	45

(b) C = (5 x 6) + 20 = 30 + 20 = **50**
(c) $20 is used to buy a skirt, so $27 is left for shirts. It costs $25 to buy 5 shirts and $30 to buy 6 shirts. She can therefore buy at most **5** shirts.

Workbook

Exercise 3, pp. 111-112 (answers p. 98)

Reinforcement

Extra Practice, Unit 9, Exercise 2, pp. 151-152

Test

Tests, Unit 9, 2A and 2B, pp. 89-95

Chapter 3 – Graphing Changes in Quantity

Objectives

♦ Complete a table for a simple linear function.
♦ Graph a simple linear function using the values from the table.
♦ Find additional values using the graph of the function.
♦ Create a table of values for points on a given line.
♦ Use the table to derive an equation for the line.

Material

♦ Coordinate grids (appendix pp. a23-a24)
♦ Rulers

Vocabulary

♦ Satisfies the equation

Notes

In this chapter, your student will relate a table showing the relationship between two quantities to a graph of a line through the points where the quantities are expressed as an ordered pair.

The table on the right shows the value of y given the value of x for the equation $y = x + 2$. If each set of quantities is written as (x, y), then each set can be graphed on a coordinate grid. If the relationship is linear (and they will be at this level), then the points can be connected with a straight line.

x	1	2	3	4	5	6
y	3	4	5	6	7	8

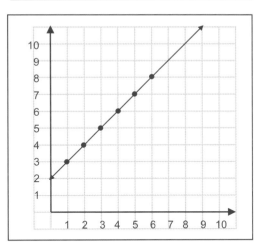

When y is expressed in terms of x in an equation such as $y = x + 2$, this indicates that the value of y depends on the value of x. So y is the dependent quantity and x is the independent quantity. On the graph the x-coordinate generally represents the independent quantity and the y-coordinate generally represents the dependent quantity. This is why the equations will always be given here as y equals some expression of x (rather than $x = y + 2$, for example).

When the graphed points are joined by a line, every point on the line **satisfies the equation** $y = x + 2$. Thus, we can use the graph to find the value of y given x by finding the point on the line where the x-coordinate is the same value as that given for x. Then the y-coordinate will be the corresponding value for y. This is a graphical solution to an equation, rather than a calculated solution. At this level, and for the rest of *Primary Mathematics*, it is easier to find a calculated solution rather than graphing the equation. Even graphing calculators, when graphing an inputted equation, use calculated values to create the graphs, and calculate the values of y for any value of x inputted.

This chapter is just a brief introduction. It is not necessary to dwell on it in depth since some of the same concepts will be taught again in *Primary Mathematics* 5B, 6A, and 6B.

For your information, equations in the form of $y = mx + b$, where m and b are constants, are called linear equations; the points (x, y) that satisfy the equation all lie on a straight line.

(1) Graph changes in quantity

Activity

Write the equation $y = x$. Have your student find y for $x = 2$, 3, and 4.

Tell your student that we can make an ordered pair with x and y. In this case, both quantities are equal. Ask him to graph the ordered pairs. Point out that the points are in line and ask him to draw a line through them and extend it. Guide him in drawing an accurate line (through the corners of the grid squares).

Indicate another place where the line crosses an intersection in the graph, such as (7, 7). Draw a point there and ask her for the corresponding ordered pair. To find the x-coordinate, we can draw or imagine a vertical line from the point (parallel to the y-axis) to the x-axis. Where it intersects the x-axis is how far the point is from the origin along the x-axis. Similarly, to find the y-coordinate, we see where a horizontal line from the point intersects the y-axis. Not surprisingly, the point has the same x-coordinate and y-coordinate, as do the other points on the line. Therefore, this point also works in the equation $y = x$. Tell her that when we have two values, one for x and one for y, that make the equation $y = x$ true, we say the values *satisfy the equation*. Show with a few other points along the line that the coordinates of the point satisfy the equation $y = x$. In fact, the coordinates for all points along the line satisfy the equation $y = x$.

Repeat with the equation $y = 2x - 1$. Create a table and have your student find the values of y for $x = 2$, 3, and 4 and write the ordered pair (x, y) for each set of values. Use the larger coordinate grid on appendix p. a24 have her graph the points. Have her find another point at an intersection, and see if the coordinates of that point also satisfy the equation.

Tell your student that since all points along the line we drew using some of the values work in the equation, we can use the line for the equation to find the value of y given any other value of x without having to put it in the equation and calculate it. Ask him to find other values of y for given values of x.

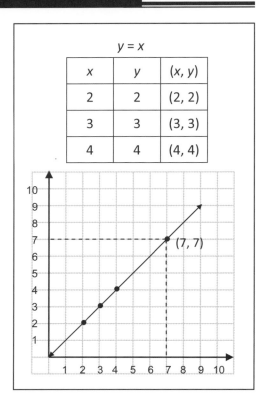

$y = x$

x	y	(x, y)
2	2	(2, 2)
3	3	(3, 3)
4	4	(4, 4)

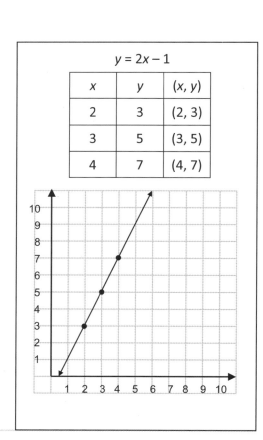

$y = 2x - 1$

x	y	(x, y)
2	3	(2, 3)
3	5	(3, 5)
4	7	(4, 7)

Discussion

Concept p. 100

Task 1, p. 101

1: In this task, instead of graphing the points from a table to create the line, the line is given and the student is asked to create a table of values from points on the line, examine the relationship between the x and y coordinates, and then derive an equation expressing this relationship. The relationship is simple; y is always one less than x. This task also shows that we can use the line to find x when given y (not just y when given x).

To find the length of a rectangle if the width is 6 cm, find the point where the first coordinate is 6, and determine the second coordinate of that point, which will the be length. The point is at (6, 8). If the width is 6 cm, then the length is 8 cm.

1. (a) $y = \mathbf{7}$ when $x = 8$
   (b) $x = \mathbf{7}$ when $y = 6$
   (c) $x = \mathbf{1}$ when $y = 0$
   (d)

x	2	3	4	5	6	7
y	1	2	3	4	5	6

   (e) $y = x - 1$

Workbook

Exercise 4, pp. 113-114 (answers p. 98)

For 2(b), you can ask your student to also draw the line.

Test

Tests, Unit 9, 3A and 3B, pp. 97-104

Workbook

Exercise 3, pp. 111-112

1. (a)

Number of steps	1	2	3	4	5	6	n
Perimeter	4	8	12	16	20	24	4 x n

(b) P = **4 x n**

(c) P = 4 x 20 = **80**

The perimeter is 80 units if the number of steps is 20.

2. (a)

P	18	18	18	18	18	18	18	18
w	1	2	3	4	5	6	7	8
l	8	7	6	5	4	3	2	1
A	8	14	18	20	20	18	14	8

(b) 20 m^2

(c) $l + w = $ **9**

(d) $l = 9 - w$

3. $n = 0.01 \times m$ or $n = m \div 100$; your student should look at the relationship between m and n for the two given values and see if it works for the other ones by substituting the values for n into the equation.

m	1	2	3	4	5		24
n	0.01	**0.02**	**0.03**	0.04	**0.05**		**0.24**

Exercise 4, pp. 113-114

1. (a)

s	1	2	3	4
p	4	8	12	16

(b) $p = 4s$

(c)

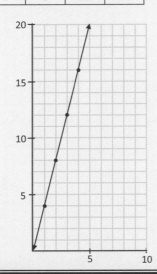

2. (a)

x	1	2	3	4	5	6
y	3	5	7	9	11	13

(b)

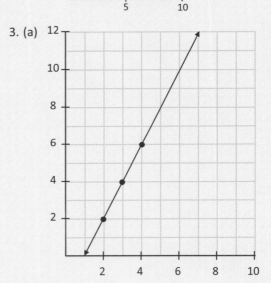

3. (a)

(b) (5, **8**)

(**6**, 10)

(**1**, 0)

Review 9

Review

Review 9, pp. 102-105

Workbook

Review 9, pp. 115-120 (answers p. 100)

Tests

Tests, Units 1-9 Cumulative Tests A and B, pp. 105-114

1. (a) 123.58, 132.85, 135.28, 251.83
   (b) 123.58: 20
       132.85: 2
       135.28: 0.2 or $\frac{2}{10}$
       251.83: 200
   (c) 123.58

2. (a) 6
   (b) 7

3. Level Negative 3

4. (a) 450.07 (b) 35.33
   (c) 30.54 (d) 107.08

5. (a) 27,360
   (b) 8262
   (c) 32,103

6. (a) $1\frac{4}{9}$ (b) $1\frac{3}{8}$ (c) $1\frac{2}{5}$
   (d) $\frac{1}{8}$ (e) 50 (f) 150

7. (a) $1\frac{9}{25}$ (b) 3.22

8. $12.25

9. (a) Perimeter = 26 cm (b) Perimeter = 62 cm
       Area = 22 cm^2 Area = 138 cm^2

10. length + width = 42 in. ÷ 2 = 21 in.
    width = 21 in. − 12 in. = **9 in.**
    The rectangle has a width of 9 in.

11. side = 5 m
    perimeter = 5 m x 4 = **20 m**
    The flower bed has a perimeter of 20 m.

12. (a) 7 h 35 min
    (b) 10:05 p.m.

13. Yes, a circle has rotational symmetry.

14. $\frac{q+p}{r}$ or $(q + p) \div r$

15. C

16. (a) yes (b) no

17.
2 units = $14
1 unit = $14 ÷ 2 = $7
3 units = $7 x 3 = **$21**
Or: $\frac{2}{5}$ of total = $14

$\frac{1}{5}$ of total = $14 ÷ 2 = $7

$\frac{3}{5}$ of total = $7 x 3 = $21

The tennis racket cost $21.

18. Cost of picture book:
    $\frac{2}{5}$ of $34 = 2 x $\frac{\$34}{5}$ = 2 x $6.80 = $13.60

    Total spent: $8.25 + $13.60 = **$21.85**
    He spent $21.85 altogether.

19. Ribbon for 6 presents:
    1.28 yd x 6 = 7.68 yd
    Total yards: 7.68 yd + 2.32 yd = **10 yd**
    She bought 10 yd of ribbon.

20. (a) (6, 3)
    (b) F
    (c) 6 units
    (d) second
    (e) 6 units
    (f) (9, **10**)
    (g)

	A	E	F
x	3	5	7
y	4	6	8

    (h) $y = x + 1$

Workbook

Review 9, pp. 115-120

1. 80,000

2. 6

3. (a) 10,000
   (b) 1000
   (c) 5

4. A: −16   B: −4   C: 8

5. (a) 67 − (100 − 52)   (b) (84 − 32) ÷ 4
       67 − 48   52 ÷ 4
       19   **13**

   (b) 72 ÷ 6 + 18 ÷ 3   (d) 47 − 28 ÷ 7 x 8
       12 + 18 ÷ 3   47 − 4 x 8
       12 + 6   47 − 32
       18   **15**

6. (a) 4.54, 5.04, 20.5, 25.4
   (b) 3.515, 5.013, 10.513, 13.015

7. 12.65

8. 1400 km

9. (a) 0.5
   (b) 3.72
   (c) 0.5

10. $\frac{13}{4}$

11. P: $3\frac{1}{4}$ Q: $3\frac{5}{8}$ R: $4\frac{1}{8}$

12. $\frac{2}{5}$

13. $1\frac{4}{5}$, $1\frac{1}{8}$, $\frac{5}{6}$, $\frac{3}{4}$

14. $\frac{20 \text{ cm}}{100 \text{ cm}} = \frac{1}{5}$

15. 98 + 98 + 153 = **349**

16. 24 x 14 = **336**

17. 25 x 4 x 12 = 100 x 12 = **1200**

18. $49.50 ÷ 3 = **$16.50**

19. Total = 0.58 km x 6 = **3.48 km**

20. Female workers = 60 − 48 = 12; $\frac{12}{60} = \frac{1}{5}$

21.

    10 qt

    5 units = 10 qt
    1 unit = 10 qt ÷ 5 = 2 qt
    8 units = 2 qt x 8 = **16 qt**

22. 10:50 a.m.

23. 2 h 45 min

24. (20¢ x 10) + (35¢ x 6) + (50¢ x 8) =
    $2.00 + $2.10 + $4.00 = **$8.10**

25.

26. Yes.   (Assume the dotted line is a line of
            Symmetry.)

27. (a)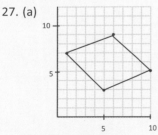

    (b) parallelogram

28. (a) (5, **9**), (5, **1**)
        (**1**, 5), (**9**, 5)
    (b) 8 units

29. b = (2 x 14) − 2 = 28 − 2 = **26**

30.
    $2290

    computer
    oven
                    ?

    1 unit = $2290 ÷ 5 = $458
    4 units = $458 x 4 = **$1832**
    The computer costs $1832 more than the oven.

31. Weight of mushrooms: 0.43 lb x 6 = **2.58 lb**
    3.05 lb − 2.58 lb = **0.47 lb**
    The empty basket weighs 0.47 lb.

Unit 10 – Data Analysis and Probability

Chapter 1 – Organizing and Analyzing Data

Objectives

- Represent data on tally charts and line plots.
- Identify the median for numerical data sets.
- Identify the mode for a categorical set of data.

Materials

- Graph paper

Vocabulary

- Survey
- Data
- Line plot
- Median
- Mode
- Tally chart

Notes

In *Primary Mathematics* 3, students learned how to present **data** using tables and bar graphs and to record data from probability experiments on tally charts. In this chapter, your student will learn how to record data from **surveys** using **tally charts** and **line plots**, and then summarize the data by looking at median and mode which are commonly used types of summary data. Students will learn about another type of summary data, mean (or average), in *Primary Mathematics* 5. Some of the data in this chapter will be recorded in a simple bar graph (textbook p. 110); students have already used bar graphs in earlier levels. More complex bar graphs will be reviewed in Chapter 4 of this unit.

The **median** is the middle value in a set of data and is representative of the data as a whole since half of the values are above it and half of the values are below it. The median is less sensitive to extreme values than the mean, or average. To find the median, we can order the values from least to greatest and mark off pairs of data starting at the ends. With an odd number of values, there is only one number in the middle. With an even number of values there are two middle numbers. If the two values in the middle are the same, the median is that value. If they are different, the median is halfway between the two values (the average of the two values).

We can also use a **line plot** to find the median (see p. 107 in the textbook). A line plot is useful for small numbers of values that are not too far apart. It also has the advantage of being able to record data directly onto it. To find the median on a line plot, mark off pairs of X's starting at the ends, from the bottom up.

The **mode** is the value that occurs most frequently in a set of data. The mode is a useful summary statistic for data that is categorical such as the favorite flavor of ice cream. It is called categorical data since it deals with categories that are not ordered or numerical. Some data, such as the age of Josh's friends on textbook p. 111, can be looked at both numerically and categorically (age as a number of years or age as a category, e.g. all 10 year olds). Both median and mean can be useful statistics for this kind of data.

Students will encounter median and mode again, as well as mean, in *Primary Mathematics* 5 and 6, so it is not necessary to spend a lot of time on this topic. It is up to you whether you want to have your student do actual surveys and collect data at this time.

(1) Find the median of a set of data

Activity

Tell your student that when we need to find out about something, we start by collecting information. The information collected is called *data*. You can use the following example, or one of your own devising. The census collects information about people's age and housing to determine how to allocate federal or state funds. To collect the data for a census, the government sends out a form for each household to fill out. In asking questions, the government is doing a *survey*. Once the data is collected, it needs to be organized and analyzed. For example, they might want to know what income is representative or typical of a group of people in a particular area, that is, the *median* income. For example, the estimated median income for Brewer, Maine, in 2007 was $44,559 (http://www.city-data.com/).

Discussion

Concept pp. 106-107

Remind your student that the marks are called tallies and the advantage of them is that we don't have to keep a running total while collecting data, we can count them later. When we have five marks, we put the fifth mark across the other four. Then we can later count the marks in fives.

Point out that the median number, 3, is representative of the number of TV sets in the home. An equal number of homes have the same or less than 3 TV sets as have the same or more than 3 TV sets.

Ask your student which is the lowest value for the data (0) and which is the highest value (5) for the data on p. 107. Point out that the median is not the middle number for the range of data, which would be halfway between 0 and 6; it is the value for the middle of the set of data when put in order. If Lila collected 2 more results (shown at the right) , 4 TV's and 6 TV's, the median would then be 4.

0, 1, 2, 2, 3, 3,④ 4, 4, 4, 4, 5, 6

Tasks 1-3, pp. 108-109

2: Since there are an even number of data values, there will not be a single value in the middle. If the two middle values are the same, that value is the median.

3: If the two middle values are different, the median is halfway between them. You can tell your student that to find a value halfway between two values, he can add half the difference between the two values to the lower value. This is illustrated by the number line (which does not relate directly to the problem; the answer to (b) is not 44).

Ask him to find the value halfway between 83 and 110.

1. (a) 77, 77, **83, 85, 86, 90, 95**
(b) 95
(c) 77
(d) 85
2. 8
3. (a) 22, 32, 35, 40, 45, 46, 50, 54, 61, 67
(b) 45.5

Halfway between 110 and 83 is 86.5:
$110 - 83 = 27$
$27 \div 2 = 13.5$
$83 + 13.5 = 96.5$

Workbook

Exercise 1, pp. 121-123 (answers p. 110)

(2) Find the mode of a set of data

Activity

Discuss ways in which stores or companies determine what products to sell and what new products to try. They don't just pick new products to put in to their stores by guessing. A store that uses a store card which you need to swipe in order to buy products at reduced price collects information on everything you buy. That way the store can determine which types of items are popular and are bought most often by what kinds of shoppers. Once the set of data is organized, the product that is most popular with a certain age group of customers would be what they liked the best. It would appear most often. There could be more than one product that appears equally the most often.

Discussion

| 4. (a) green |
| (b) blue |

Task 4, p. 109

Be sure your student understands the definition of *mode*. The mode is not the same as the median, although it is sometimes the same value. Have him look again at the *line plot* on p. 107 and find the mode. Although 4 TV's is the mode, because it appears most often, it is different from the median, which is 3. The mode is not as representative of the group as a whole, since there are only 4 families with 4 TV's, but there are 6 families with less than 4 TV's.

Tasks 5-6, pp. 110-111

| 5. (a) 18 |
| (b) 4 |
| (c) 2 |
| (d) dogs and cats |
| (e) birds |
| 6. (a) 11 |
| (b) 8 years |
| (c) 11 years |
| (d) 3 years |
| (e) 9 years |
| (f) 9 years |
| (g) 9 years |

5: We can see the mode easily with bar graphs, not just line plots. Bar graphs are more useful for data with larger numerical values than line plots. Ask your student how *tally charts*, bar graphs, and line plots make it easier to determine the mode than a table that just shows the numbers would be.

Practice

Ask your student to find both the median and mode of the following sets of data:

⇒ 1, 1, 2, 3, 5, 8, 1, 3, 2, 1   median: 2; mode: 1

⇒ 95, 98, 96, 99, 98, 59, 59, 80, 68, 98   median: 95.5; mode: 98

Workbook

Exercise 2, p. 124 (answers p. 110)

Reinforcement

Extra Practice, Unit 10, Exercise 1, pp. 157-158

Test

Tests, Unit 10, 1A and 1B, pp. 115-121

Reinforcement

Ask your student to come up with a survey question that interests him, perform a survey (collect data), organize the data in a table, line plot, or simple bar graph, and determine the median and mode.

Chapter 2 – Probability Experiments

Objectives

♦ Conduct probability experiments and express the outcomes verbally and numerically.

Material

♦ Coins
♦ Dice or number cubes

Vocabulary

♦ Event
♦ Outcome

Notes

In *Primary Mathematics* 3A, students conducted simple probability experiments to determine if different possible **outcomes** of an **event** are likely, unlikely, or impossible, and made predictions based on their results. At that level, they did not assign a numerical value to the outcome.

In this chapter, your student will extend this understanding of probability to representing the likelihood of an outcome from experimental data as a fractions.

In *Primary Mathematics* 6B, students will learn to calculate the probability of equally likely outcomes for an event mathematically (theoretical probability).

Probability theory is a branch of mathematics concerned with the analysis of random phenomena. Probability theory was first used in attempts to analyze games of chance in the sixteenth century. The theory has become a powerful and widely applicable branch of mathematics and has widespread use in business, science, and industry. It is a foundation of statistics, is essential to many human activities that involve quantitative analysis of large sets of data, and applies to complex systems such as quantum mechanics.

Probability deals with predicting how likely it is that each possible outcome will happen. One way to determine how likely it is that something will occur is to take a statistical approach. This means devising an experiment and repeating the experiment over and over again many times. If you repeat the experiment enough times, you will get results that are very close to the probability you can calculate mathematically in situations where it is possible to find the theoretical probability. Statistical or experimental probability, for example, is how insurance companies determine risk and so how much premium to charge.

The probability of an outcome is usually given as a number from 0 through 1 expressed as a fraction, decimal, or percent. If the probability of an outcome is 0, it is impossible. If an outcome is certain, it has a probability of 1. The more unlikely an outcome is, the closer its probability is to 0. The more likely an outcome is, the closer its probability is to 1.

The likelihood of outcomes only helps determine what will probably happen, but does not determine what will actually happen.

(1) Use fractions to represent probable outcomes

Discussion

Concept p. 112

> They got heads **6** out of 10 times.

Tell your student that sometimes we want to be able to predict whether something is likely to happen, or determine how probable it is. One way to predict how likely something will happen in the future is to compare how likely it was in the past. If it usually snows in December, then the probability that it will snow on any particular day in December is high. The more data we have showing that it snowed in December, the more confidence we have that it might snow on a given day in December.

To get the likelihood that something will happen, we can do probability experiments. For example, if we toss a coin, which is more likely, heads or tails? In this experiment, the *event* is the coin toss. There are two possible *outcomes* for the *event*: that the coin lands heads up or that it lands tails up. We can represent the number times each outcome occurs as a fraction of the total number of outcomes. Since the coin was tossed 10 times, the total is 10. Since heads resulted 6 out of the ten times, we can say that $\frac{6}{10}$ or $\frac{3}{5}$ of results are heads; that is, heads landed up $\frac{3}{5}$ of the time. Based on these results alone, neither Mary's nor Sam's predictions seem to be correct and heads is slightly more likely than tails.

Activity

Have your student repeat this experiment but toss the coin (on a cloth or towel) more times, such as 50 times and record the outcomes in either a tally chart or line plot. Then have him determine what fraction of the outcomes are heads and tails, simplifying the fraction if possible. He should find that the outcomes are now much closer to one half for each. Sam's prediction in the textbook makes sense, because with a fair coin it is just as likely to land on either side, and heads is one out of two sides, or $\frac{1}{2}$ of the sides. If we tossed the coin only once, and got tails, then we could not really say that we get tails all the time from that one outcome. In order to determine the probability of an outcome using experimental data, we need to collect a lot of data, not just from one or ten throws. However, we cannot ever accurately predict the outcome of any single outcome unless the probability is 0 or 1. We can't know that the coin *will* land heads up. We have a half of a chance of being correct.

Discussion

Tasks 1-2, p. 113

Ask your student if she thinks that the values in Task 1 will remain the same if she threw the die a lot more times than 18. They would not. Ask her what she thinks the probability of each outcome should be, with enough throws. Since there are 6 possible outcomes, then the value should be close to $\frac{1}{6}$ for each outcome.

1. (a) 18 (b) **4 out of 18**
   (c) $\frac{2}{9}$ (d) **3 out of 18**
   (e) $\frac{3}{18} = \frac{1}{6}$ (f) $\frac{2}{18} = \frac{1}{9}$
   (g) $\frac{9}{18} = \frac{1}{2}$ (h) $\frac{12}{18} = \frac{2}{3}$

2. (a) **8 out of 20** (b) $\frac{2}{5}$
   (c) **12 out of 20** (d) $\frac{3}{5}$

Activity

You can ask your student to repeat this experiment but roll the die more often and see if the fraction for each separate outcome gets closer to $\frac{1}{6}$. You could also ask him to predict what fraction of the time he would roll an odd number or an even number, or any number greater than a number such as 4, and use the results from many rolls to test his prediction.

Workbook

Exercise 3, p. 125-126 (answers p. 110)

Reinforcement

Extra Practice, Unit 10, Exercise 2, pp. 159-160

Test

Tests, Unit 10, 2A and 2B, pp. 123-130

Enrichment

Use a bottle cap. Write H for heads on a square of masking tape and stick it on top of the bottle cap. Write T for tails on another square of masking tape and stick it inside the bottle cap. Have your student toss the bottle cap 50 times on a towel and record the results. Are the results close to one half for each side, as with a coin? If not, ask her to explain why not. The bottle cap may be a biased, or unfair "coin," in which case one side is favored over the other. If the results are somewhat close to one half for each side, but one side seems to be favored, ask her how she could be more certain that the cap is or is not biased. Note that in the first activity on p. 112, it would seem that the coin was biased. But with more throws, if your student did the activity, the results indicate that the coin is less likely to appear to be biased. Would more throws determine whether the bottle cap is biased? (yes)

Chapter 3 – Order of Outcomes

Objectives

♦ Represent all the possible outcomes of a probability situation using a tree diagram.

Material

♦ Coins
♦ Dice
♦ Permutation Tree (appendix p. a26)
♦ Combination Tree (appendix p. a27)

Vocabulary

♦ Tree diagram

Notes

In this chapter, your student will learn how to use a **tree diagram** to determine the number of possible outcomes for a probability situation.

When we use probability, we are usually looking at an event that has equally likely outcomes. For example, provided we use a fair coin, there are two equally likely outcomes when we toss a coin, heads or tails. The set of all possible outcomes is called the sample space. The sample space for tossing a coin could be expressed as (H, T). The sample space for tossing two coins would be (HH, HT, TH, TT).

In *Primary Mathematics* 6, students will learn to compute theoretical probability of independent events, such as getting two heads if a coin is tossed twice. To do so, they need to know how many possible outcomes there are.

Later, your student will learn how to find the total number of possible outcomes in a sample space using multiplication. It is not necessary for him to compute possible results by multiplication at this stage, only to represent the different possibilities by using a tree diagram.

If a coin is tossed three times, as on p. 114, the order is significant. There are 8 possible outcomes when order is significant. If, on the other hand, rather than three separate events of one coin toss, there is a single event of three coins tossed at the same time, there is no order for heads or tails, and there are therefore only 4 separate outcomes. Since this chapter is an introduction to finding the number of outcomes for independent events, order is important. This guide does include some activities and optional enrichment for investigating the number of outcomes to simple situations if order is not significant. There is one question on the Test A for this chapter asking students to make groups out of the outcomes in order to determine the number of possible outcomes when order is not significant.

This chapter is not about combinatorics. The 2009 printing of the tests have some errors in the answers for Test B due to confusion between the purpose of this chapter (to find the number of possible outcomes in a probability experiment using a tree diagram where order is important) and finding the number of permutations (order is important) versus combinations (order is not important). If you are using the tests, change all terms that say "combination" to "outcome" on Test B for problems 1-4 and treat problem 5 as "extra credit" since it goes beyond the concepts of this chapter (the answer can be found using a modified tree diagram). The answers for 2 and 3 on test B are then both C.

(1) Find all possible outcomes

Activity

Ask your student to list all the possible outcomes for the following situations (she can use abbreviations):

⇒ Answers on at true-false question   (T, F)

⇒ Tossing a coin   (H, T)

⇒ Rolling a 6-sided die   (1, 2, 3, 4, 5, 6)

⇒ Drawing a ball out of a bag that has a red, blue, yellow, and green ball   (r, b, y, g)

⇒ Drawing two balls out of a bag that has a blue, yellow, and green ball   (by, bg, yg)

⇒ The sex of two children in a family, from oldest to youngest   (BB, BG, GB,GG)

⇒ Tossing a coin and rolling a 6-sided die   (H1, T1, H2, T2, H3, T3, H4, T4, H5, T5, H6, T6)

⇒ Tossing two coins   (HH, HT, TT)

⇒ Tossing a coin twice, in order   (HH, HT, TH, TT)

⇒ Tossing a coin three times, in order

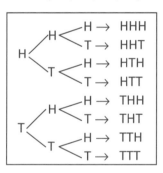

Let your student start listing the possible outcomes for the last. There are eight of them. Then tell him that we need a systematic way to list the outcomes. One way is to draw what is called a *tree diagram*. Guide him in developing the tree. We would start with two branches for the first toss, heads or tails (H or T). Then, ask him how many possibilities there are for the second toss if the first toss is heads. There are two, so there are two branches from the H. Continue guiding her in drawing the rest of the "tree". Once we have shown the third toss, then we just follow the path to list the results for each toss.

The final diagram shows that there are 8 possible outcomes if the order is important. Ask your student for the number of possible outcomes if three coins were tossed at the same time. There would be only 4, all heads, all tails, two heads and one tail, and two tails and one head. HHT, HTH, and THH are the same, and HTT, THT, and TTH are the same.

Discussion

Concept p. 114
Task 1, p. 115

You can ask your student how many outcomes there would be if two different spinners were used, or ask how many possible outcomes are there if there are 2 red, 2 green, and 2 blue socks in a drawer and you pick out two socks without looking. There are 6 (both red, both blue, both green, one red and one green, one blue and one red, one blue and one green).

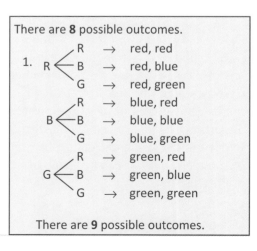

There are **8** possible outcomes.

1. R	R	→	red, red
	B	→	red, blue
	G	→	red, green
B	R	→	blue, red
	B	→	blue, blue
	G	→	blue, green
G	R	→	green, red
	B	→	green, blue
	G	→	green, green

There are **9** possible outcomes.

Workbook

Exercise 4, p. 127-128 (answers p. 110)

Reinforcement

Extra Practice, Unit 10, Exercise 3, pp. 161-162

Test

Tests, Unit 10, 3A and 3B, pp. 131-138

Enrichment

Use a copy of the Permutation Tree on appendix p. a26 and one coin. Have your student look at the tree. It lists all the possible outcomes for tossing a coin four times. Ask her to predict what fraction of the time she will get each outcome. Have her toss a coin 4 times, record the result for each toss, and then make a tally mark on the chart for the results of that experiment. She should perform the experiment 64 times and make a tally mark on the chart at the end of the appropriate branch each time. Then she should write each outcome as a fraction of a whole and simplify if possible.

Now give your student a copy of the Combination Tree on appendix p. a27 and four coins. In this tree, the order of the outcome is not significant. Ask him to predict what fraction of the time he will get each outcome. Then have him toss the four coins 64 times and tally the results as above.

Ask your student compare and discuss the results for the two experiments. For example, if we add the fractions in the first experiment that show three heads and one tail (HHHT, HHTH, HTHH, THHH) we should get a fraction close to that for 3 heads and 1 tail in the second experiment.

Workbook

Exercise 1, pp. 121-123

1. (a) 8, 8, 9, **9, 9, 9, 9, 10, 10, 10, 11, 11, 12**
   (b) 8 years
   (c) 12 years
   (d) 12 − 8 = 4 years
   (e) 9 years

2.

	×			×
	×			×
×	×		×	×
10	11	12	13	14

   (a) 10
   (b) 14
   (c) 12

3. (a) 15
   (b) 5
   (c) 0
   (d) 5 − 0 = 5
   (e) 1

4. (a) 138, 143, 145, 149, 150, 152
   (b) 147 cm

Exercise 2, p. 124

1. (a) blue
   (b) black
   (c) blue

2. (a) cars
   (b) cars
   (c) motorcycles
   (d) 15
   (e) 18

Exercise 3, pp. 125-126

1. (a) 5
   (b) $\frac{5}{20} = \frac{1}{4}$
   (c) **4** out of **20**
   (d) $\frac{4}{20} = \frac{1}{5}$
   (e) **7** out of **20**
   (f) $\frac{7}{20}$
   (g) **4** out of **20**
   (h) $\frac{4}{20} = \frac{1}{5}$

2. (a) 24
   (b) **4** out of **24**
   (c) $\frac{1}{6}$
   (d) **5** out of **24**
   (e) $\frac{5}{24}$
   (f) **3** out of **24**
   (g) $\frac{1}{8}$
   (c) $\frac{1}{3}$

Exercise 4, pp. 127-128

1.
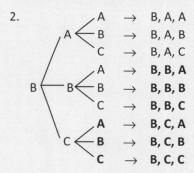

H   H →   Heads, heads
     T →   Heads, **tails**

T   H →   **Tails, heads**
     T →   **Tails, tails**

There are **4** possible outcomes.

2.
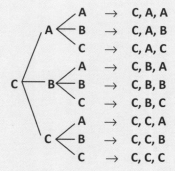

	A	A →	B, A, A
	A	B →	B, A, B
		C →	B, A, C
		A →	**B, B, A**
B	B	B →	**B, B, B**
		C →	**B, B, C**
		A →	B, C, A
	C	B →	B, C, B
		C →	B, C, C
		A →	C, A, A
	A	B →	C, A, B
		C →	C, A, C
		A →	C, B, A
C	B	B →	C, B, B
		C →	C, B, C
		A →	C, C, A
	C	B →	C, C, B
		C →	C, C, C

There are 27 total possible outcomes.

Chapter 4 – Bar Graphs

Objectives

♦ Interpret bar graphs.

Notes

In *Primary Mathematics* 3 and 4 students learned to draw and interpret bar graphs. This chapter is a primarily a review. Your student will also learn how to interpret bar graphs with two bars for each category.

Bar graphs are generally used to display categorical data such as months of the week or methods of transport. In *Primary Mathematics* 5, students will also learn how to present categorical data using pie charts (or pie graphs).

A bar graph has two axes, one for the categories and one for the numerical data for how many items are in each category. The categories can be on either the vertical or the horizontal axes; in this review they are on the horizontal axis but students have seen bar graphs in earlier levels where the categories were on the vertical axis. The numerical axis is marked with units and gives the scale of the graph. Students have had ample practice in determining the scale of number lines in earlier levels of *Primary Mathematics*.

Creating bar graphs from data is not included in this lesson. If you want to have your student collect data and draw bar graphs as part of this lesson, you will have to help him determine what type of data to collect and guide him in determining a good scale for the graph paper depending on the data and labeling the graph appropriately. The scale of the graph should be drawn accurately in order for the graph to be interpreted correctly; that is, the distance for each unit should be consistent. However, there are likely plenty of opportunities to have your student draw bar graphs in other subjects, such as science, and you can use this lesson simply to remind and familiarize him with this method of displaying data.

You can also teach your student how to use computer software to create the bar graph, but practice with paper and pencil initially will result in a better understanding of bar graphs than just using software.

You can have your student look at bar graphs in newspaper or online articles. Pay particular attention to how the graph might be drawn in a way to bias the data to cause the reader to agree with a given conclusion, such as having a larger scale so that differences are more pronounced when in actuality the numbers are close to each other and not significantly different. Or the scale might be condensed to minimize differences, even though numbers for each category are quite different. Also, sometimes part of the scale is left out or the scale is not consistent for the entire graph. When interpreting bar graphs it is important to pay careful attention to the numbers and the scale, not just the height of the bars.

(1) Interpret bar graphs

Discussion

Concept p. 116

Discuss the relationship between the data in the table and the data on the bar graph. Tell your student that bar graphs are often used for data that is in categories, in this case days of the week. Refer back to the simple bar graph on p. 110. In that bar graph, the categories were the types of pets.

> On **Saturday**, there were 3 times as many visitors as on Monday.
>
> There were at least twice as many visitors (at least 300) for **4** days.

Remind your student that the number lines on the two sides of the bar graph are called axes. Point out that the vertical axis is scaled – each major division represents 100 visitors. Ask what the minor divisions (between the hundreds) represent. (50 visitors)

Ask your student what information the bar graph gives. It shows the relationship between the number of visitors and days of a week in a way that makes it easy to compare the number of visitors on different days, since it is easy to compare the height of the bars. Ask her to answer the two questions, and tell whether they used the table or the graph to answer the question.

Ask your student for circumstances in which we might prefer to look at data in a table versus a bar graph. For example, it is easier to know the exact data value from the table, but it is easier to see relationships between the data on the bar graph.

Remind your student that bar graphs can be oriented so that the categories are on the vertical axis, in which case we would compare the lengths of the bars, rather the height, in looking for differences between categories.

Tasks 1-2, pp. 117-118

2: This graph shows two bars for each category, one for boys and one for girls. Ask your student why such a graph might be useful. We can use it to compare not only the boys to girls in each group, but also either the boys or the girls between groups. For example, if the groups were different activities at the community center, we could see that all four activities have a similar attendance, i.e. no one activity is particularly unpopular or popular, but that boys tend to prefer the activity represented by B more than girls do.

> 1. (a) 150
>   (b) 250
>   (c) 1050
>   (d) $\frac{200}{1050} = \frac{4}{21}$
>
> 2. (a) Groups B and C
>   (b) Group D
>   (c) 42
>   (d) 173

Workbook

Exercise 5, pp. 129-130 (answers p. 118)

Reinforcement

Extra Practice, Unit 10, Exercise 4, pp. 163-164

Test

Tests, Unit 10, 4A and 4B, pp. 139-145

Chapter 5 – Line Graphs

Objectives

♦ Represent data in a line graph.
♦ Interpret line graphs.
♦ Interpret conversion graphs.

Material

♦ Graph paper
♦ Appendix p. a24
♦ Appendix p. a28

Vocabulary

♦ Line graph
♦ Conversion graph

Notes

In earlier levels of *Primary Mathematics*, students learned to interpret picture graphs, tables, bar graphs, and line plots. Here, your student will learn about line graphs as another form of data representation.

The **line graphs** in this section display data measurements collected over a period of time. Points are plotted similar to a coordinate graph, with time (the independent variable) on the x-axis (horizontal axis) and amount (the dependent variable) on the y-axis (vertical axis). The data points are connected by lines drawn between successive data points.

Line graphs are generally used to see if there are any trends in data over time. From the line graph, we can immediately see the increase or decrease, over time, in the value of the measured quantity. For example, a graph that shows how the height of a boy changes over time would trend upwards until a certain age, when it would flatten out. There might be steeper parts on the graph where there was a "growth spurt."

In the linear coordinate graphs created by substituting values in an equation that your student learned about in the previous unit, points along the line between the data points were valid points that satisfied the equation being graphed.

With line graphs created from data, the lines connecting the successive data points are for convenience only, and are not valid data points. The dots indicate which points are actual measurements. There is no way of knowing whether a point along the line accurately reflects the situation at that particular point.

Line graphs are also sometimes used for **conversion graphs**. These display a specific linked relationship between two quantities in the same way as a coordinate graph does, such as the exchange rate between Singapore dollars and U.S. dollars shown on p. 121 of the textbook. Unlike a line graph composed of individually measured data points, the points shown on the conversion graph are the points used to draw the graph, but in this case the line joining the points shows the exact values for any data point. As with the linear coordinate graphs, once we have the line we can find the value along the vertical axis for any value along the horizontal axis and vice versa.

(1) Interpret line graphs

Discussion

Concept p. 119

(a) Increase: 3500 – 2500 = **1000**
(b) Decrease: 3500 – 2000 = **1500**
(c) Difference: 4000 – 3500 = **500**

Discuss the relationship between the data in the table and on the line graph. The data could have been presented in a bar graph, but instead of drawing bars the top of the bar is essentially marked with a dot and then the dots are connected with a line.

Tell your student that line graphs are created similarly to the coordinate graphs he learned earlier, except that earlier he only graphed points that resulted in straight lines. For this particular graph, instead of numbers along the bottom, the names of the months are used (but we could also have used the date, or the numbers for each month of the year). The month tells us how far to go to the right to graph the point, and the number of people is how far we go up to graph the point. Point out the scale on the vertical axis. For coordinate graphs, the scale is usually 1 unit for 1. Here, the marked units are for 1000. Ask him what the value of the unmarked units are, i.e. where the horizontal lines cross the vertical axis but there are no labels. They are for 500, 1500, 2500, and 3500. Make sure your student can "read" the points on the graph, i.e., assign values or ordered pairs to them. For example the first point is for (Aug, 2500) and corresponds to the first column of data in the table.

Tasks 1-2, p. 120

1. (a) Increase: 200 – 125 = **75**
   (b) Decrease: 200 – 100 = **100**

2. (a) Increase: $200,000 – $100,000 = **$100,000**
   (b) Decrease: $200,000 – $50,000 = **$150,000**

Tell your student that the bar graphs and line plots (and pictographs) she has learned so far are used mainly to make comparisons between data for different categories, such as which is larger and which is smaller. Line graphs are used more to look at trends in the data, such as when the numbers increase over a period of time. As you discuss the two graphs on this page, discuss any trends that might be observed from them, and what the results might mean to the company.

1: The number of people in the supermarket peaks at 8 p.m., possibly because earlier is dinnertime and later is bedtime. So the supermarket has data to justify the number of people to have on the cash registers, for example, at a particular time.

2: The number of sales peaks at the third week. This could be because during the first and second week some of the people who come to the trade show are "shopping around" or deciding whether they really want to buy the products. Then, if they do decide, they might want to hurry up and buy it before the end of the show or the products are gone.

Tell your student that in these two graphs, the data on the vertical axis is numerical. With line graphs, we usually put the independent data on the horizontal axis. The time is independent of the number of people whereas the number of people is dependent on the time.

Workbook

Exercise 6, pp. 131-134 (answers p. 118)

Enrichment

Have your student look at a copy of appendix p. a28. This is the same information as that in the graph on p. 119 of the textbook, but with different scales. Ask him which graph would be used by someone who wanted to show that there was a considerable increase in attendance in December compared to August and so the pool hours should be reduced in August and which graph would be used by someone who wanted the pool hours to be the same for all 5 months. (If he wonders why there may be more swimmers later in the year, you could suggest that this graph reflects the climate in Singapore where winters are warmer than summers, or that this is an indoor pool.) With the second graph, we could be making the point that the attendance does not change much from month to month and so the pool should have the same hours each month. Both graphs, however, contain the same data.

Have your student look at various line graphs in newspaper articles, scientific journals, and textbooks. Relate the data in the line graph with the conclusions stated in the article. Tell her that when a graph of any kind is being used to make a point, it is essential for the reader to pay careful attention to the scale of the graph. The scale used can make a graph misleading, even if the data are correct. Misleading graphs are often used in advertisement and politics.

Have your student look at other graphs that might be of interest. There is a lot of data on the internet. For example, there are sites that give weather information (high and low temperature, precipitation, number of sunny days, etc.) for different cities. It is easy to then compare different areas of the world with respect to weather. These sites often have various types of tables and graphs, including bar graphs and pie charts, and so are good places to discuss the benefits of different ways to present data.

These lessons do not include specific instructions on gathering, recording, and graphing data. There should be plenty of opportunities for that in other subjects, such as science or social studies. If not, you can provide opportunities for actually collecting and graphing data. Try to pick something that your student is interested in. If nothing comes to mind, data on weather, such as high and low temperatures each day (resulting in a graph with two lines) is a good option. Guide him in determining the scale that needs to be used on each axis depending on the graphing paper. This can be tricky. You can tell him to draw the axes of the graph first. Label the horizontal axis for the independent variable and the vertical axis for the dependent variable. For each, determine the highest value. Then count the squares along each axis. Divide the highest value by the number of squares to get a rough estimate of a good scale. It is often a good idea to adjust that so there are 2, 4, or 5 squares between each unit that will be marked so that it is easy to place points between those units. For example, if the numbers marked on the vertical axis are 0, 100, 200, 300, etc., it is better to have two or five squares between 0 and 100, depending on how small the squares are on the graph, rather than 3, so that it is easier to estimate where to place a point in between 100 and 200, for example. He might have to lengthen an axis a bit from the original drawing to fit the values in, or shorten it, to have a scale that is easy to use.

You can also teach your student how to make graphs using computer software, so that the data can be presented in a slideshow or a nicely formatted printed report. Most spreadsheet software have a graphing component.

(2) Conversion graphs

Discussion

Task 3, p. 121

Tell your student that the graph shown on this page shows another way line graphs can be used. This graph shows the relationship between Singapore dollars and U.S. dollars. It is called a *conversion graph*, because it shows how to convert U.S. dollars to Singapore dollars, or vice-versa.

> 3. (a) 18 Singapore dollars
>   (b) 5 U.S. dollars
>   (c) 8 U.S. dollars
>   (d) $u = \frac{1}{2} \times s$
>
> $s = 2 \times u$
>   (e) 60 Singapore dollars

Ask your student if this reminds them of any other kind of graph. This type of line graph is very similar to the coordinate graphs where they graphed the relationship between two points when that relationship resulted in a straight line.

3(d): If necessary, have your student create a table in order to see the relationship between the two quantities. Have your student find the formulas for both u as a function of s and s as a function of u.

3(e): Ask her which equation she found for 3(e) is most useful when given the number of U.S. dollars.

You can ask your student to find some other values not on the chart using the two equations, including some that include cents.

After your student has answered the questions for this task, ask him whether a bank would be more likely to use a graph or an equation such as he found for Task 3(d). Nowadays, with calculators and exchange rates changing each day, they would use the equations, or formulas, showing the relationship and simply plug in, or substitute, the given value for either U.S. or Singapore dollars to find the other value.

Workbook

Exercise 7, pp. 135-136 (answers p. 118)

Reinforcement

Extra Practice, Unit 10, Exercise 5, pp. 165-166

Test

Tests, Unit 10, 5A and 5B, pp. 147-157

Enrichment

Have your student look at the scales on the axes for the graph on p. 121. She should notice that the scale is different for each axis. Two squares are used to mark 1 for the horizontal axis and 1 square for the vertical axis.

Provide your student with a copy of the coordinate graph on appendix p. a24 and have him graph the same points on it. If necessary, he can first create a table. He can also extend the line beyond (10,20). Then have him compare the two graphs and note any differences.

Graphs of both kinds (different scales and the same scale) are shown at the right.

Ask your student to compare the two graphs. Both contain the same information, but in the second graph the line is steeper. Ask her which graph presents a more accurate visual representation of the data and why. The second graph does, because it shows the line going up faster than across. It goes up two squares for every square it goes across. Each U.S. dollar is worth two Singapore dollars, so each time the U.S. dollars goes up one, the Singapore dollars goes up two.

Tell your student that whenever the two values we want to graph are close, such as both within a similar range, or when graphing simple equations as in the previous unit on coordinate graphs, we should make the axes the same, if possible, since that makes it easier to see the relationship between the two numbers.

In other situations, such as data collected over time in months, hours, or minutes, as in the line graphs on p. 119-120 of the textbook, the two types of data are not the same (e.g. number of people versus time) and it is not possible to have the two axes the same. Then the scale depends on the range of the data and how we want it to fit on the page.

(Later, in the secondary level, your student will compare graphs such as $y = x$, $y = mx$, and $y = mx + b$ and see how m and b change the slope of the line. To see how changing m and b transforms the line, the axes have to be the same.)

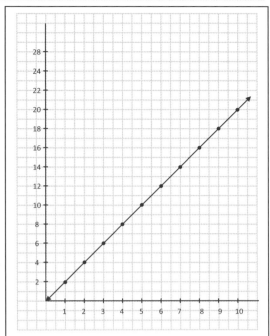

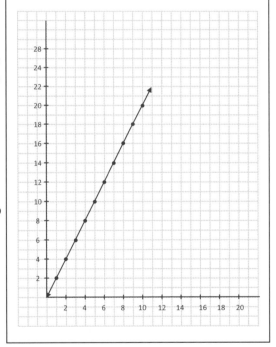

Workbook

Exercise 5, pp. 129-130

1. (a) January
   (b) June
   (c) 18
   (d) 11
   (e) 78

2. (a) A: $18,000, B: $15,000
   (b) Sunday
   (c) $16,000
   (d) $\frac{17}{104}$

Exercise 6, pp. 131-134

1. (a) 200
   (b) 2001-2002
   (c) 700
   (d) 3400

2. (a) Wednesday
   (b) 375
   (c) Saturday
   (d) 75
   (e) Tuesday to Wednesday

3. (a) 3 cm
   (b) 4 cm
   (c) Tuesday to Wednesday
   (d) Thursday to Friday; 4 cm
   (e) 4 days

4. (a) 7 a.m.
   (b) 130
   (c) 8 a.m. to 9 a.m.
   (d) 7 a.m. to 8 a.m.
   (e) 9 a.m. to 10 a.m.

Exercise 7, pp. 135-136

1. (a)

Singapore $	1	2	3	4	5
Hong Kong $	4	8	12	16	20

   (b) $2.50
   (c) $18
   (d) $s = \frac{h}{4}$
   (e) $25

2. (a) 3 min
   (b) 4.5 min
   (c) 40 ℓ
   (d) 70 ℓ
   (e) (i)

Time (min)	1	2	3	4	5
Volume (ℓ)	20	40	60	80	100

   (ii) $V = 20 \times t$

Review 10

Review

Review 10, pp. 122-127

5(d): Assume Celsius temperature scale. If your student assumes Fahrenheit scale, then the answer is +72 degrees.

16: Estimates may vary.

27: Your student may have to trace the figure and extend the lines. Do not expect measurements to be exactly $\angle x = 140°$ and $\angle y = 130°$ due to errors inherent in tracing, extending lines, and accuracy of the protractor, but they should be close.

28: Rather than writing in the textbook, have your student indicate the pair of lines that are perpendicular and the pair that are parallel. Have him use a set-square and ruler; there are some intersections that look like they are right angles but are not quite.

31: Your student can draw a tree diagram.

Remember that this guide will not show every possible method, just one or sometimes two suggested solutions. Your student may draw a model for more problems than are shown in this guide, or may not need to draw a model for a problem where one is shown in this guide, or the model may be different. Set your own guidelines for when to require model drawing depending on your student's abilities and needs.

Workbook

Review 10, pp. 137-143 (answers pp. 121-122)

Tests

Tests, Units 1-10 Cumulative Tests A and B, pp. 159-170

---

1. P: 89,100   Q: 89,800   R: 90,400

2. (a) 600,000
   (b) 4,991,300
   (c) 6,294,500

3. ten thousand

4. (a) −12
   (b) −8

5. (a) −15 ft   (b) −9 s
   (c) +5 lb   (d) +40 degrees
   (e) −12 m

6. $22.50

7. $\frac{2}{100}$ or 0.02

8. 5

9. (a) 21
   (b) 48.02
   (c) 0.06
   (d) 0.2

10. 10.03

11. 40.26, 40.62, 42.06, 42.6

12. 3.75

13. 45.96 + 68.2 = **114.16**

14. 30.05 − 9.2 = **20.85**

15. 14 x 35 = **490**

16. (a) 14.8   (b) 382.8   (c) 139.65

17. (a) 1.2   (b) 3.5   (c) 2.7

18. 8 out of 16 = $\frac{1}{2}$

19. $1\frac{3}{5}$

20. $1 - \frac{3}{8} - \frac{3}{8} = \frac{2}{8} = \frac{1}{4}$

   Ken had $\frac{1}{4}$ of the pizza.

21. $\frac{1}{3}$ of 120 = $\frac{120}{3}$ = **40**

   There were 40 children.

   (Continued next page.)

22.

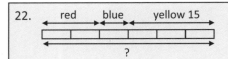

red   blue   yellow 15

?

(a) $\frac{1}{2}$ of the marbles were yellow.

(b) 3 units = 15
    6 units = 15 x 2 = **30**
    There were 30 marbles altogether.

23. 3 weeks: $10.50
    9 weeks is 3 times 3 weeks.
    9 weeks: 3 x $10.50 = **$31.50**
    Or
    1 week = $10.50 ÷ 3 = $3.50
    9 weeks = $3.50 x 9 = $31.50
    She will save $31.50 in 9 weeks.

24. (a) Side: **8 in.** (since 8 x 8 = 64)
    (b) Perimeter: 8 in. x 4 = **32 in.**

25. Divide into a square and a rectangle:
    (6 cm x 6 cm) + (4 cm x 9 cm)
    = 36 cm^2 + 36 cm^2
    = **72 cm^2**
    Or enclose in a rectangle:
    (10 cm x 9 cm) – (6 cm x 1 cm)
    – (6 cm x 2 cm)
    = 90 cm^2 – 6 cm^2 – 12 cm^2
    = 72 cm^2

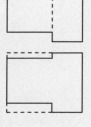

26. 2

27. $\angle x = 140°$, $\angle y = 130°$

28.

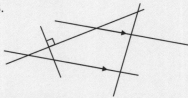

29. (a) (4, 0)
    (b) (7, 5)
    (c) F
    (d) H

30. (a) 128 cm, 130 cm, 130 cm, 135 cm, 140 cm
    (b) 130 cm
    (c) 12 cm

31. (a) 24
    (b) 6

32. Length of AD and BC:
    36 cm – (2 x 5 cm) = 26 cm
    BC = 26 cm ÷ 2 = **13 cm**
    The length of rectangle ABCD = 13 cm.

33. (a) 16
    (b) 65
    (c) $20,000

Workbook

Review 10, pp. 137-143

1. (a) 79,031
   (b) 55,100
   (c) 23.29
   (d) 18.21

2. (a) 1
   (b) 9

3. (a) $35,500
   (b) 8 m
   (c) 17 yd

4. (a) $\frac{2}{3}$; 1

   (b) 3.15, 3.35
   (c) 4, 0, −4, −8

5. 3.05

6. 0.4

7. $\frac{32}{40} = \frac{4}{5}$

8. $2

9. Check drawing.

10. Check drawing.

11.

    No. (no rotational symmetry)

12. 6

13.
```
        1   →   1 + 1 = 2
  1 <   2   →   1 + 2 = 3
        3   →   1 + 3 = 4
        1   →   2 + 1 = 3
  2 <   2   →   2 + 2 = 4
        3   →   2 + 3 = 5 ✓
        1   →   3 + 1 = 4
  3 <   2   →   3 + 2 = 5 ✓
        3   →   3 + 3 = 6 ✓
```
3 outcomes have sums greater than 4.

14. $\frac{5}{7}$

15. $2\frac{2}{5}$

16. (a) 6.66
    (b) 0.27
    (c) 24
    (d) 0.55

17. **9**   There are 3 thirds in 1, so there are 3 x 3 = 9 thirds in 3.

18. 4.3

19.
```
                        $1800
              TV  [  |  |  |  ]  } ?
          Camera  [  ]
```
4 units = $1800
1 unit = $1800 ÷ 4 = $450
Total = $1800 + $450 = **$2250**
or total = 5 units = $450 x 5 = $2250
The total cost is $2250.

20. Cost of shirts: $12.50 x 2 = $25.00
    Total money: $39.85 + $25.00 = **$64.85**
    He had $64.85 at first.

21. (a) 16
    (b) $\frac{4}{16} = \frac{1}{4}$

    (c) $\frac{2}{16} = \frac{1}{8}$

    (d) $\frac{8}{16} = \frac{1}{2}$

    (e) $\frac{6}{16} = \frac{3}{8}$

22. (a)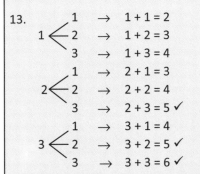

    (b) 36

(Continued next page.)

Workbook

23.

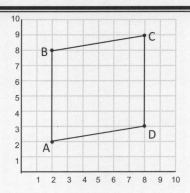

(a) (8, 3)
(b) BC and AD, or AB and DC
(c) 6 units
(d) second coordinates or *y*-coordinates

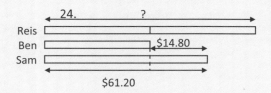

24.

1 unit = $61.20 - $14.80 = $46.40
2 units = $46.40 x 2 = **$92.80**
Reis received $92.80.

25. Width: 54 cm^2 ÷ 9 cm = 6 cm
    Length + width: 6 cm + 9 cm = 15 cm
    Perimeter: 15 cm x 2 = **30 cm**
    The perimeter of the rectangle is 30 cm.

26. Amount left after making dress:
    4.5 ft − 0.9 ft = 3.6 ft
    Amount for each cushion:
    3.6 ft ÷ 5 = **0.72 ft**
    She used 0.72 ft to make each cushion.

27.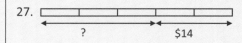

1 unit = $14 ÷ 2 = $7
3 units = $7 x 3 = **$21**
The racket cost $21.

Unit 11 – Measures and Volume

Chapter 1 – Adding and Subtracting Measures

Objectives

♦ Review conversion units for measurement.
♦ Convert between measurements within a measurement system.
♦ Review addition and subtraction of measures in compound units.

Material

♦ Meter stick

Vocabulary

♦ Conversion factor
♦ Compound units

Notes

In *Primary Mathematics* 3B, students learned to add and subtract measurements in **compound units**. This is reviewed in this chapter. Measurement in compound units are those that involve two units, such giving the length of an object in meters and centimeters, or in feet and inches. Adding and subtracting measures in compound units allows a student to both practice measurement unit conversions and mental math.

There are two distinct systems of measurement, the metric system and what is called the U.S. customary system in this curriculum. The metric system has become the global language of measurement and is used by about 95% of the world's population. It is used in science in the U.S. The metric system is based on powers of 10. If this is your student's first exposure to the metric system, you should go back to *Primary Mathematics* 3B, including the Home Instructor's Guide, and do the units on measurement there in order to become familiar with the metric system. There is not adequate review in this chapter to really understand or get a "feel" for the metric system of measurement.

In the metric system, it is easy to convert between units, so two measurements in, for example, meters and centimeters can be easily converted to centimeters only and then added together. However, adding and subtracting by making the next larger unit or taking from the next larger unit is very similar to using mental math strategies. So measures in compound units can be added or subtracted by first adding the values for the larger units, and then the values for the smaller units using mental math strategies for making 100 or 1000. Students less adept at mental math can simply convert or "rename" 1 of the larger units.

4 m 25 cm + 5 m 90 cm: add the meters: 9 m; add the cm: 1 m 15 cm; answer is 10 m 15 cm.

In the U.S. customary system, it is not as easy to convert between units and the conversion factor varies. 1 foot is 12 inches, but 1 yard is 3 feet. So it makes more sense to first add the larger units together, and then the smaller ones. Then, we only have to convert one of the larger units. Or, we can apply mental math strategies similar to those learned for base-10. This will give your student some experience working with other bases than base-10, such as "making a 3" or "making a 12."

3 ft 10 in. + 5 ft 7 in.: add the ft: 8 ft.; add the inches: 1 ft 5 in.; answer is 9 ft 5 in.

(1) Convert measures

Discussion

Top of p. 129

Rather than immediately looking at the chart in the textbook, you may want to determine what conversion factors your student remembers. You may also want to review the abbreviations. Measurements are no longer abbreviated with periods even for the U.S. customary system, except for the abbreviation for inches, in., to distinguish it from the word in. Abbreviations are not standardized, and though this curriculum uses a cursive ℓ for liters some will use a lowercase non-cursive l and some uppercase L. Different abbreviations may be used for time, such as "s" or "sec" for seconds. In this table, the time measurements are often written out.

Make sure your student understands that there are two measurement systems shown here for length, weight, and capacity. Point out that for the metric system the conversion for kilometers to meters is 1000, for kilograms to grams is 1000, and for liters to milliliters is 1000. However, for meters to centimeters it is 100. Tell him that if he knows what the prefixes mean, that will help him remember whether to use 1000 or 100. Centi- means 100 and is easy to remember since it sounds like cents, and 1 dollar is 100 cents. 1 meter is 100 *centi*meters. Kilo - means 1000, and is used for the larger measurement. So 1 *kilo*meter is 1000 meters, and 1 *kilo*gram is 1000 grams. Milli- also means 1000 but is used for the smaller measurement, so 1 liter = 1000 *milli*liters.

Point out that the conversions in the table at the top of the page go from the larger unit to the smaller. 1 meter is the larger unit, and 1 centimeter is the smaller unit. 100 is called the *conversion factor* for meters to centimeters. To convert between a larger unit to a smaller one we can simply multiply by the number of smaller units in the larger unit of measurement. If 1 m is 100 cm, then 2 m is 2 x 100 cm, and 134 m is 134 x 100 cm.

> 1 m = 100 cm
> 2 m = 2 x 100 = 200 cm
> 34 m = 34 x 100 cm = 3400 cm

Task 1, p. 129

Point out that it can help to write or mentally think of the conversion factor for 1 of the units of measurement as the girl is doing in her thought bubble. Since 1 ft is 12 inches, then 4 ft is 4 times 12 inches. This is similar to using the part-whole model for multiplication in word problems where we have the value of 1 unit and need to find the value of more units.

> 1. (a) 1 ft = 12 in.
>   4 ft = 4 x 12 in. = **48** in.
> (b) 1 m = 100 cm
>   9 m = 9 x 100 = **900** cm
> (c) 1 day = 24 hours
>   8 days = 8 x 24 hours = **192** hours
> (d) 1 lb = 16 oz
>   12 lb = 12 x 16 oz = **192** oz

The similarity in the numerical portion of the answers to (c) and (d) could lead to a discussion on how to solve these problems mentally using factors or number bonds. For example, 8 x 24 is (8 x 20) + (8 x 4), or 24 x 2 x 2 x 2.

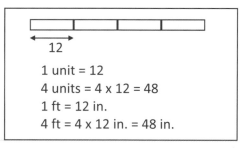

> 1 unit = 12
> 4 units = 4 x 12 = 48
> 1 ft = 12 in.
> 4 ft = 4 x 12 in. = 48 in.

Task 2, p. 129

Remind your student that measurements can be made in *compound units*. For example, if the capacity of a jug is more than 4 liters but less than 5 liters, the amount between 4 and 5 liters can be measured in milliliters. In these problems, we want to convert the compound measurement to the measurement in just the smaller unit. To do so with 4 ℓ 250 ml, all we need to convert is the part that is 4 ℓ, and then add that to 250 ml.

Even though the steps are shown in the solutions at the right, your student can solve the problems mentally.

Task 3, p. 129

Before your student does this task, discuss converting from a smaller unit to a larger unit. Ask her to find the number of feet in 48 inches. Here, we are essentially grouping the inches by 12's, since 1 group of 12 inches is 1 ft. So we divide. 48 inches divided into groups of 12 gives 4 groups, so there are 4 ft in 48 inches. Now ask her to find the number of feet in 50 inches. When we divide by 12, there will be 2 inches left over. We can think of this as splitting 50 into the number of inches that would make a whole number of feet, and the rest. There are not an exact number of feet in 50 inches and we express the answer as a compound unit; feet and the left-over inches.

Notice how much easier it is to convert the metric measurements mentally than the U.S. customary ones.

Workbook

Exercise 1, problems 1-3, pp. 144-145 (answers p. 135)

---

2. (a) 4 ℓ 250 ml
   = (4 x 1000) ml + 250 ml
   = 4000 ml + 250 ml = **4250 ml**
   (b) 5 km 40 m
   = (5 x 1000) m + 40 m
   = 5000 m + 40 m = **5040 m**
   (c) 4 years 5 months
   = (4 x 12) months + 5 months
   = 48 months + 5 months
   = **53 months**
   (d) 1 hour 20 minutes
   = (1 x 60) minutes + 20 minutes
   = 60 minutes + 20 minutes
   = **80 minutes**

---

48 in. = __ ft
Since 12 in. = 1 ft, group by 12
48 ÷ 12 = 4
48 in. = 4 ft

50 in. = ? ft
50 ÷ 12 = 4 R 2
50 in. = 4 ft 2 in.

$$50 \text{ ft}$$
$$\diagup \quad \diagdown$$
48 in.   2 in.
= 4 ft

---

3. (a) 3 ft = 1 yd
   8 ft = **2** yd **2** ft (8 ÷ 3 = 2 R 2)
   $\diagup \diagdown$
   6 2
   (b) 100 cm = 1 m
   602 cm = **6** m **2** cm
   $\diagup \diagdown$
   600 2
   (c) 1000 g = 1 kg
   2400 g = **2** kg **400** g
   $\diagup \diagdown$
   2000 400

Enrichment

List the conversions for time. Point out that the list is in order from the largest unit of measurement, year, to the smallest one, second. Ask your student how he could use the list to determine conversion factors for other times, such as finding how many seconds there are in a day.

Since there are 24 hours in a day, and 60 minutes in an hour, then there are 60 x 24 minutes in a day. Similarly, there are 60 x 60 x 24 seconds, or 86,400 seconds, in a day. We multiply the conversion factors for days to hours, hours to minutes, and minutes to seconds together to get one for days to seconds.

Point out that we cannot use this chart alone to find the number of seconds in a year, since we cannot go from months to weeks. If we wanted to find the number of seconds in a non leap-year, we can use 1 year = 365 days. Then the number of seconds in a year is 365 x 24 x 60 x 60 = 31,536,000 seconds (about thirty-one and a half million seconds).

We can use this same concept to find the number of cups in a gallon or quart. If we add in a half-gallon, then we can just multiply by 2's for all conversions with U.S. customary units for capacity.

1 year = 12 months
1 week = 7 days
1 day = 24 hours
1 hour = 60 minutes
1 minute = 60 seconds
1 day = 24 x 60 x 60 seconds   = 86,400 seconds
1 year = 365 days (non leap-year)
1 year = 365 x 86,400 seconds   = 31,536,000 seconds

1 gallon = 2 half-gallons
1 half-gallon = 2 quarts
1 quart = 2 pints
1 pint = 2 cups
1 gallon = 2 x 2 x 2 x 2 cups   = 16 cups

Your student may be interested in other units of measurement in the metric system. She should already know that there are 1000 millimeters in a meter. She can research other units between millimeter and kilometer, such as hectometer, decameter, and decimeter. She may also want to research smaller units of measurements, such as picometers, nanometers, or angstroms. There are also units of measurement for many other things than length, weight, and capacity, such as paper sheet sizes, steel metal gauges, tennis racket gauges, hurricane intensity, solar flare intensity, and others.

(2) Add and subtract in compound units

Activity

Write the problems at the right and have your student do the subtractions. To subtract, he needs to convert the larger measurement to the smaller. Point out that we can use mental math strategies for "making 100" or "making 1000" with the metric system measurements. We can "make" other numbers with the other measurements, like "make 12" when subtracting inches from feet.

1 m – 42 cm = _____ cm	(58 cm)
1 km – 390 m = _____ m	(610 m)
1 ℓ – 7 ml = _____ ml	(993 ml)
1 ft – 7 in. = _____ in.	(5 in.)
1 yd – 2 ft = _____ ft	(1 ft)
1 lb – 7 oz = _____ oz	(9 oz)
1 h – 15 min = _____ min	(45 min)

Write the second set of problems and have your student do the subtractions. Rather than convert all of the larger measurements into the smaller ones, she just needs to convert one of them. With the U.S. customary units, it would be a pain to convert 19 ft to 228 in., subtract 3 in., and then convert back using division. It is much easier to split 19 ft into 18 ft and 1 ft and subtract 3 in. from the 1 ft. For the last two examples, we can subtract from the larger unit, then add back in the smaller unit, or "rename" the larger unit, both strategies she has learned with whole numbers.

4 m – 25 cm = _____ m _____ cm	(3 m 75 cm)
10 km – 90 m = _____ km _____ m	(9 km 910 m)
3 ℓ – 985 ml = _____ ℓ _____ ml	(2 ℓ 15 ml)
19 ft – 3 in. = _____ ft _____ in.	(18 ft 9 in.)
8 lb – 10 oz = _____ lb _____ oz	(7 lb 6 oz)
5 gal – 3 qt = _____ gal _____ qt	(4 gal 1 qt)
4 h 10 min – 25 min = ___ h ___ min	(3 h 45 min)
(4 h – 25 min + 10 min)	
(3 h 70 min – 25 min)	
19 ft 1 in. – 3 in. = _____ ft _____ in.	(18 ft 10 in.)
(19 ft – 3 in. + 1 in.)	
(18 ft 13 in. – 3 in.)	

Write the third set of problems at the right and have your student do the additions. He can simply add and then convert. Or, he can add mentally by "making" the conversion factor. This is shown with the number bonds. In the example at the right for adding meters, it might be easier to simply add, rather than trying to make 1000. If your student is good at mental math, it should be easy to make 12, 3, 4, or 16 with the U.S. customary measurements. He can use either method, but encourage him to use mental math strategies when feasible.

65 cm + 40 cm = _____ m _____ cm	(1 m 5 cm)
/\	
5 60	
780 m + 390 m = _____ km _____ m	(1 km 170 m)
/\	
220 170	
400 ml + 750 ml = _____ ℓ _____ ml	(1 ℓ 150 ml)
/\	
600 150	
9 in. + 6 in. = _____ ft _____ in.	(1 ft 3 in.)
/\	
3 3	
8 oz + 10 oz = _____ lb _____ oz	(1 lb 2 oz)
/\	
2 6	
3 ft 7 in. + 10 in. = _____ ft _____ in.	(4 ft 5 in.)
/\	
5 2	
8 h 30 min + 20 min = ___ h ___ min	(8 h 50 min)

Discussion

Concept p. 128

When finding the total weight, we first add the larger units together (kilograms or pounds) and then the smaller units. This greatly simplifies the computation, especially with U.S. customary measurement.

Discuss different methods that could be used to add the final units for each of these examples.

In the first example the text shows the girl simply adding 450 g and 650 g to get 1 kg 100 g, which is then added to the 5 kg, resulting in another kilogram. Thus the answer is 6 kg 100 g. We could also "make the next 1000" by splitting the 650 g into 550 g and 100 g.

In the second example, the text shows the boy subtracting 650 g from 1 kg and then adding back in the remaining 450 g. We could also rename 1 kg 450 g as 1450 g and subtract 650 g from that.

In the third example, the boy adds the ounces together and then converts, resulting in one more pound. We could also split 8 oz into 2 oz and 6 oz and add by making the next pound.

In the fourth example, the boy subtracts 14 oz from one of the pounds. We could also rename 2 lb 10 oz as 1 lb 26 oz and subtract 12 oz from that.

Practice

Tasks 4-7, p. 130

Workbook

Exercise 1, p. 145, problem 4 (answers p. 135)

Reinforcement

Extra Practice, Unit 11, Exercise 1, pp. 169-170

Test

Tests, Unit 11, 1A and 1B, pp. 171-174

---

4. (a) 450 g   (b) 5 in.
   (c) 40 min   (d) 5 oz

5. (a) 37 ft 7 in.
   (b) 34 kg 620 g
   (c) 12 min 0 s
   (d) 8 gal 2 qt

6. 20 lb – 13 lb 9 oz = **6 lb 7 oz**
   The smaller watermelon weighs 6 lb 7 oz.

7. 1 ℓ 450 ml + 650 ml + 1 ℓ 20 ml = **3 ℓ 120 ml**
   The three beakers contain a total of 3 ℓ 120 ml of solution.

Chapter 2 – Multiplying Measures

Objectives

♦ Multiply measures in compound units by a 1-digit number.

Material

♦ Mental Math 19 (appendix p. a7)

Notes

In this chapter, your student will learn to multiply a measurement in compound units by multiplying each unit separately and then adding the products.

Students have learned that in order to multiply a multi-digit number, such as 485, by a 1-digit number, such as 8, the digit in each place-value is multiplied by 8 and then the products are added together, renaming as needed.

485 x 8 = (400 x 8) + (80 x 8) + (5 x 8) = 3200 + 640 + 40 = 3880

(This process is simplified by using the standard algorithm for multiplication.)

Similarly, measurements in compound units can be multiplied by a 1-digit number by multiplying the value for each unit of measurement separately and then adding the products together, converting as needed.

4 kg 850 g x 8 = (4 kg x 8) + (850 g x 8) = 32 kg + 6800 g = 32 kg + 6 kg + 800 g = 38 kg 800 g

4 ft 8 in. x 8 = (4 ft x 8) + (8 in. x 8) = 32 ft + 64 in. = 32 ft + 5 ft 4 in. = 37 ft 4 in.

With the metric system, it is just as easy to convert to the smaller unit and multiply (using the standard multiplication algorithm or mental math strategies) and then convert back.

4 kg 850 g x 8 = 4850 g x 8 = 38,800 g = 38 kg 800 g

We could do this also with the U.S. customary system, but that often results in more computation steps or dividing a larger number to convert back.

4 ft 8 in. x 8 = 56 in. x 8 = 448 in. = 37 ft 4 in.

(4 ft x 12 + 8 in. = 48 in. + 8 in. = 56 in.)   (448 ÷ 12 = 37 R 4)

(1) Multiply measures

Discussion

Concept p. 131

Remind your student that multiplication can be thought of as repeated addition. We could add the weight of the three packages together. To do this, we would add the kilograms together, add the grams together, and then add both sums together. Essentially we are multiplying the kilogram by 3, the 200 grams by 3, and adding the products together.

Tasks 1-3, p. 132

1: This time, when we multiply the smaller unit by 4, we get more than 1 kilometer. So before we add it to the product of the kilometers and 4, we need to convert it to kilometers and meters.

The answers at the right show the steps in solving the problems, but you should allow your student to find the answer mentally; do not require him to write down all these steps.

Practice

Write the following expressions and have your student solve them.

⇒ 6 h 30 min x 6   (39 h)
(This can be done either by 30 min x 6 = 180 min = 3 h, or by recognizing that 30 min x 2 = 1 h, so 30 min x 6 = 3 h.)

⇒ 2 ft 10 in. x 5   (14 ft 2 in.)

⇒ 3 lb 8 oz x 4   (14 lb)
(Since 8 oz is half of a pound, 4 of them is 2 lb.)

⇒ 16 gal 3 qt x 6   (100 gal 2 qt)

⇒ 4 kg 201 g x 5   (21 kg 5 g)

Workbook

Exercise 2, pp. 146-147 (answers p. 135)

Reinforcement

Extra Practice, Unit 11, Exercise 2, pp. 171-174

Mental Math 19

Test

Tests, Unit 11, 2A and 2B, pp. 175-180

---

3 kg **600** g
3 kg **600** g

```
  1 kg 200 g          1 kg 200 g
  1 kg 200 g          /      \
+ 1 kg 200 g        1 kg     200 g
  3 kg 600 g        x 3      x 3
                    3 kg     600 g
                      \      /
                    3 kg 600 g
```

1. 1 km 300 m x 4
= (1 km x 4) + (300 m x 4)
= 4 km + 1200 m
= 4 km + 1 km 200 m
= **5** km **200** m
5 km **200** m

2. 2 gal 3 qt x 4
= (2 gal x 4) + (3 qt x 4)
= **8** gal **12** qt
= 8 gal + 3 gal   (12 ÷ 4 = 3)
= **11** gal
11 gal

3. 4 m 25 cm x 3
= (4 m x 3) + (25 cm x 3)
= **12** m **75** cm
2 x (4 m 25 cm + **12 m 75 cm**)
= 2 x **17 m**
= **34 m**
Or: 2 x 4 x (4 m 25 cm) = (8 x 4 m) + (8 x 25 cm) = 32 m + 2 m = 34 m

Chapter 3 – Dividing Measures

Objectives

♦ Divide measures in compound units by a 1-digit number.

Notes

In this chapter, your student will learn how to divide measures in compound units.

To divide measures in compound units, we again split the measure into the larger and smaller unit and divide each one separately. Since division of the larger unit will sometimes result in a remainder, we need to combine the remainder with the smaller unit, and then divide. This process will allow students to both review division and see how the concept behind the division algorithm (divide the digit in the largest place-value first, rename the remainder and add to the digit in the next place value and divide that, etc.) applies to other situations than division of whole numbers. At the secondary level, they will use the same algorithm to divide polynomial equations.

For example, to divide 7 hours and 10 minutes by 6, we first divide the hours by 6, get a remainder of 1 hour, rename it to minutes, add this remainder to the 10 minutes, and then divide the minutes by 6. This process is similar to the division algorithm, but since the number system is base-60, not base-10 as with whole numbers, we can't simply "drop" the 3 down next to the remainder of 1 hour. We have to first convert the hour to 60 minutes and add that to the 30 minutes.

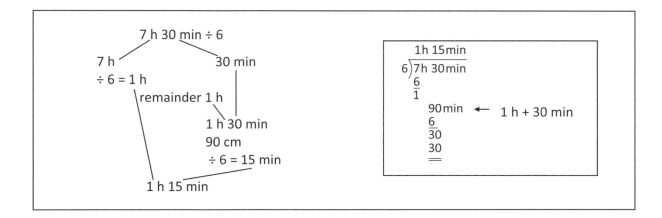

(1) Divide measures

Discussion

Concept p. 133

Tell your student that we can also divide measures in compound units by dividing each unit separately, but with division we have to deal with remainders when a number does not divide evenly.

In this example, we first divide 5 meters by 4. This leaves a remainder of 1 meter. So we have 1 meter and 20 centimeters that still need to be divided. Ask your student how we can divide the remaining amount. We can convert the remainder of 1 m to 100 centimeters and combine it with 20 centimeters. We can then divide 120 centimeters by 4.

Optional: You can show your student how this process is similar to division of whole numbers. To divide 520 by 4, we first divide the hundreds and get a remainder of 1 hundred which is renamed as 10 tens and added to the 2 tens. Then 12 tens are divided by 4. Similarly, the remainder from the division of the larger unit is "renamed" and added to the smaller unit. (With centimeters and meters this works just the same as with hundreds and ones, since there are ten tens of centimeters in a meter.)

Tasks 1-3, p. 134

Guide your student through the steps for solving these division problems.

1: The larger unit can be evenly divided by 5, so we just divide the smaller unit as well by 5.

2: Dividing 7 h by 6 gives 1 h with a remainder of 1 h, or 60 min. That is added to the 30 min and then the total minutes divided by 6.

3: In this case, we cannot divide 3 liters by 8 and get a whole liter, so all 3 liters are converted to milliliters before the total milliliters are divided.

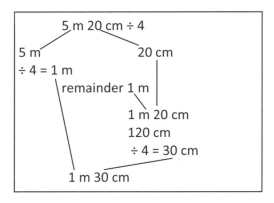

1 m 30 cm
1 m 30 cm

$5 \text{ m } 20 \text{ cm} \div 4$

5 m
$\div 4 = 1 \text{ m}$
remainder 1 m

20 cm

1 m 20 cm
120 cm
$\div 4 = 30 \text{ cm}$

1 m 30 cm

$$\begin{array}{r} 1\text{m } 30\text{ cm} \\ 4\overline{)5\text{m } 20\text{ cm}} \\ \underline{4} \\ 1 \end{array}$$
$\qquad$ 120 cm $\leftarrow 1 \text{ m} + 20 \text{ cm}$
$\qquad$ $\underline{120}$

1. 5 kg 650 g ÷ 5 = ?
   5 kg ÷ 5 = 1 kg
         650 g ÷ 5 = 130 g
   5 kg 650 g ÷ 5 = 1 kg **130** g
   1 kg 130 g

2. 7 h 30 min ÷ 6 = ?
   7 h ÷ 6 = 1 h remainder 1 h (60 min)
       60 min + 30 min = 90 min
       90 min ÷ 6 = 15 min
   7 h 30 min ÷ 6 = **1 h 15** min
   1 h 15 min

3. 3 ℓ 200 ml ÷ 8 = ?
       3 ℓ 200 ml = 3200 ml
       3200 ml ÷ 8 = 400 ml
   3 ℓ 200 ml ÷ 8 = **400** ml
   400 ml

Optional: Use Task 2 to show how the process relates to the division algorithm. Here, the remainder of 1 hour is renamed as 60 minutes and then added to the 30 minutes.

```
      1h 15min
6)7h 30min
  6
  1
      90min  ← 1 h + 30 min
      6
      30
      30
      ==
```

Practice

Ask your student to solve the following problems. The division algorithm representation is shown here and will sometimes be shown in the solutions in this guide since it is a concise way to show how the problem can be solved.

⇒ 10 min 12 s ÷ 6   (1 min 42 s)

⇒ 20 lb 2 oz ÷ 7   (2 lb 14 oz)

```
    1min 45s              2lb 14oz
6)10min 12s           7)20lb  2oz
  6                     14
  4                      6
      252s                  98oz
      24                    7
      12                    28
      12                    28
      ==                    ==
```

⇒ A piece of rope 22 ft long needs to be cut into 3 pieces. The second piece has to be twice as long as the first piece and the third piece has to be three times as long as the first piece. How long will the second piece be?

(The second piece will be 7 ft 4 in. long.)

Workbook

Exercise 3, pp. 148-149 (answers p. 135)

Reinforcement

Extra Practice, Unit 11, Exercise 3, pp. 175-178

Total units = 6
1 unit = 22 ft ÷ 6 = 3 ft
            with remainder 4 ft
            4 ft = 48 in.
            48 in. ÷ 6 = 8 in.
1 unit = 3 ft 8 in.
2 units = 3 ft 8 in. x 2
            = 6 ft 16 in.
            = 7 ft 4 in.
Or
The second rope is 2 out of 6
units, or $\frac{1}{3}$ of the total.
2 units = 22 ft ÷ 3 = 7 ft
            with remainder 1 ft
            12 in. ÷ 3 = 4 in.
2 units = 7 ft 4 in.

(2) Practice

Practice

Practice A, pp. 135-136

Test

Tests, Unit 11, 3A and 3B, pp. 181-186

---

1. (a) 3 km 200 m x 5
   = 15 km 1000 m
   = **16 km**

   (b) 4 ℓ 300 ml x 4
   = 16 ℓ 1200 ml
   = **17 ℓ 200 ml**

   (c) 2 h 20 min x 5
   = 10 h 100 min
   = **11 h 40 min**

   (d) 5 kg 200 g x 3
   = 15 kg 600 g
   = **15 kg 600 g**

   (e) 6 m 20 cm x 6
   = 36 m 120 cm
   = **37 m 20 cm**

   (f) 3 yd 2 ft x 7
   = 21 yd 14 ft
   = **25 yd 2 ft**

2. (a) 2 ℓ 240 ml ÷ 2
   = **1 ℓ 120 ml**

   (b)
   ```
         2 km 650 m
   2) 5 km 300 m
      4
      1
              1300 m
              12
               100
               100
   ```

   (c) 1 h 30 min ÷ 5
   = 90 min ÷ 5
   = **18 min**

   (d)
   ```
        1 kg 500 g
   3) 4 kg 500 g
      3
      1
            1500 g
            1500
   ```

   (e) 2 m 60 cm ÷ 4
   = 260 cm ÷ 4
   = **65 cm**

   (f)
   ```
        1 ft  5 in.
   3) 4 ft 3 in.
      3
      1
          15 in.
          15
   ```

3. 1 ℓ 275 ml x 2 = **2 ℓ 550 ml**
   She used 2 ℓ 550 ml of syrup.

4. 3 kg 570 g ÷ 3 = **1 kg 190 g**
   The beans in each bag weighed 1 kg 190 g.

5. 3 h 30 min x 5 = 15 h 150 min = **17 h 30 min**
   He spent 17 h 30 min painting his house.

6. (a) 1 kg 800 g x 3 = 3 kg 2400 g = **5 kg 400 g**
   The watermelon weighs 5 kg 400 g.
   (b) 5 kg 400 g + 1 kg 800 g = **7 kg 200 g**
   The total weight is 7 kg 200 g.

7. (a) 8 h 30 min x 6 = 48 h 180 min = **51 h**
   In 6 days she works 51 hours.
   (b) 51 x $5 = **$255**
   She earns $255.

8.

3 units = 3 m 66 cm
1 unit = 3 m 66 cm ÷ 3 = 1 m 22 cm
2 units = 1 m 22 cm x 2 = **2 m 44 cm**
The longer piece was 2 m 44 cm long.

9. Total sugar: 1 kg 240 g + 1 kg 160 g = 2 kg 400 g
   2 kg 400 g ÷ 8 = 2400 g ÷ 8 = **300 g**
   She used 300 g of sugar for each cake.

10. Total ribbon: 3 m 50 cm x 2 = 6 m 100 cm = 7 m
    7 x $4 = **$28**
    She paid $28 for the ribbon.

Workbook

Exercise 1, pp. 144-145

1. (a) 2500 cm
   (b) 120 in.
   (c) 8 qt
   (d) 3000 m
   (e) 80 oz
   (f) 4000 g
   (g) 6000 ml
   (h) 264 hr

2. (a) 66 months
   (b) 6020 m
   (c) 8100 ml
   (d) 63 in.
   (e) 127 oz
   (f) 4500 g
   (g) 155 s
   (h) 3 pt

3. (a) 2 years 6 months
   (b) 1 m 1 cm
   (c) 1 h 10 min
   (d) 1 lb 14 oz
   (e) 11 yd 0 ft

4. (a) 10 hr 5 min
   (b) 1 kg 410 g
   (c) 9 ft 11 in.
   (d) 7 gal 1 qt
   (e) 17 lb 0 oz

Exercise 2, pp. 146-147

1. (a) 12 m 80 cm
   (b) 255 cm
       2 m 55 cm
   (c) 6 m 255 cm
       8 m 55 cm

2. (a) 10 ℓ 750 ml
   (b) 1600 ml
       1 ℓ 600 ml
   (c) 12 ℓ 1600 ml
       13 ℓ 600 ml

3. (a) 24 ft 8 in.
   (b) 54 in.
       4 ft 6 in.
   (c) 60 ft 54 in.
       64 ft 6 in.

4. 1 ℓ 500 ml x 3 = 3 ℓ 1500 ml = **4 ℓ 500 ml**
   The bucket holds 4 ℓ 500 ml.

5. 1 h 40 min x 4 = 4 h 160 min = **6 h 40 min**
   It takes 6 h 40 min to wash 4 loads.

6. 12 lb 12 oz x 6 = 72 lb 72 oz = **76 lb 8 oz**
   The total weight is 76 lb 8 oz.

Exercise 3, pp. 148-149

1. (a) 2 km 125 m
   (b) 400 m
   (c) 1 km 400 m

2. (a) 2 h 15 min
   (b) 20 min
   (c) 1 h 20 min

3. 6 lb 12 oz ÷ 9 = 108 oz ÷ 9 = **12 oz**
   The mushrooms in each box weigh 12 oz.

4. (a) 4 m 50 cm ÷ 3 = 3 m 150 cm ÷ 3 = **1 m 50 cm**
      Each piece of wire was 1 m 50 cm long.
   (b) 1 m 50 cm x 2 = 2 m 100 cm = **3 m**
      He used 3 m of wire to repair his toy.

5. Weight of books: 6 kg 850 g − 600 g = 6 kg 250 g
   6 kg 250 g ÷ 5 = 5 kg 1250 g ÷ 5 = **1 kg 250 g**
   Each book weighs 1 kg 250 g.

Chapter 4 – Cubic Units

Objectives

- Find the volume of solids made up of unit cubes.
- Find the volume of solids made up of 1-centimeter cubes.
- Find the volume of 2-dimensional representations of solids made up of unit cubes.

Materials

- Multilink cubes or other cubes
- Centimeter cubes
- Clay
- Rulers, meter sticks
- Appendix pp. a13, a29-a31

Vocabulary

- Volume

Notes

In *Primary Mathematics* 3B, students learned that **volume** is the amount of space a solid occupies, and that volume is measured in cubic units. They learned how to find the volume of a solid made up of unit cubes drawn in two dimensions. This is reviewed in this chapter.

In *Primary Mathematics* 3B, the term square centimeter was used for an area that is equivalent to a square 1 centimeter on each side, and cubic centimeter was used for a volume that is equivalent to a cube 1 centimeter on each side. In *Primary Mathematics* 4A, students learned to represent a unit of area with a superscript, e.g., cm^2. In this chapter, your student will learn to represent a unit of volume with a superscript 3, e.g., cm^3.

Your student should be able to visualize a solid drawn in 2 dimensions as a 3-dimensional object. The drawing might include cubes he can't see. If he has any difficulty, spend extra time having him construct the solids with actual cubes. He can use unit cubes of any size to build the solid, but he should be familiar with the size of a centimeter cube since this measure is used frequently and later he will learn that 1000 cubic centimeters is a liter. The unit cubes from a base-10 set are approximately 1-centimeter cubes.

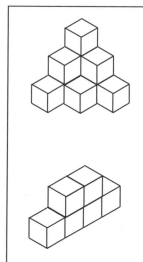

In all of these exercises, assume that the figures drawn on paper can be constructed from blocks that do not link. That is, when a block is not on the lowest level, one or more blocks have to be under it so that no block is suspended in the air. For example, in the pyramid shape at the right, there are blocks that are hidden behind other blocks. This figure needs to have 10 blocks in order to construct it, 4 of which are hidden in the drawing but need to be there to support other blocks. Also assume that there are hidden blocks *only* when one is necessary to support another block. For example, in the 6-block figure at the right, while there could be a block hiding behind the second to the last block in the bottom row, such a block would not be a supporting block, so we know the figure has just the 6 blocks that we see.

Relating 2-dimensional drawings of cubes to 3-dimensional solids can be difficult for some students. If it is difficult for your student, you may want to revisit Unit 13 of *Primary Mathematics* 3B.

(1) Find the volume of solids

Activity

Show your student a solid cube. Ask her to define *volume*. It is the amount of space a solid takes up. Length is measured in units, area is measured in square units, and volume is measured in cubic units. As with length and area, we use standard units to measure volume. So volume can be measured in cubic centimeters, cubic inches, and so on. A cubic centimeter is a cube that is a centimeter on each side. When we don't need to specify a measurement unit, we can simply use the term "unit" and a cubic unit is a cube that is a unit on each side.

Build an object out of the cubes. You can use multilink cubes and tell your student we are only measuring to volume of the cube part of it, so we are ignoring the parts that stick out or in to link them together. Have him count the cubes to find the volume.

Tell your student that we use the cubic units for volumes of objects not made out of cubes as well. When we say an object has a volume of 1000 cubic centimeters, for example, then the amount of space it takes up is the same as the amount of space 1000 centimeter cubes take up even if the shape of the space is not the same. One way to illustrate this is to use some string to cut a cube of clay to a particular size corresponding to a square made out of unit cubes, and then change its shape. Changing the shape does not change the volume.

Discussion

Concept page 137

> A: 5 cubic units   B: 9 cubic units
> C: 18 cubic units   D: 12 cubic units
> **C** has the greatest volume.

If necessary, have your student build these solids with cubes and relate the actual solid with the two-dimensional drawing.

Discuss the strategies your student used to find the number of cubes for C and D. She may have simply imagined the lower layers and counted the hidden cubes, or multiplied the number of cubes in one layer by the number of layers.

Tasks 1-3, pp. 138-139

> 1. The volume of the solid is **3** cm^3.
>   **6** solids were used to build the next solid.
>   The volume of the solid is **6** cm^3.
> 2. 32 cm^3
> 3. A: 4 cm^3 B: 12 cm^3
>   C: 10 cm^3 D: 12 cm^3

1: This task emphasizes the abbreviation used to show cubic units.

2: Ask your student how he found the volume. We could count the cubes in the top layer twice, or we could count the cubes along each side, multiply them together to get the number of cubes in the top layer (4 x 4 = 16) and double that to get the number of cubes in both layers (16 x 2 = 32).

3: You can also ask your student which solid has the largest volume and what the difference in volume is between several of the solids. Ask her how she can find the difference in volume without finding the actual volumes of the two solids. For example, to get from solid C to solid D we could move one cube down to the bottom layer and add 2 cubes on the end, so 2 cubes were added and the difference in volume 2 cm^3.

Practice

Have your student find the volume of the solids on appendix pp. a29-a30, assuming that they are made up of centimeter cubes.

Workbook

Exercise 4, p. 150 (answers p. 147)

Reinforcement

Have your student draw some solids using the isometric dot paper on appendix p. a13. He could first build the solids with cubes.

Extra Practice, Unit 11, Exercise 4, pp. 179-180

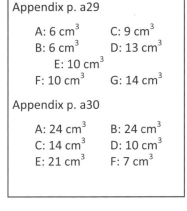

Appendix p. a29

A: 6 cm^3	C: 9 cm^3
B: 6 cm^3	D: 13 cm^3
E: 10 cm^3	
F: 10 cm^3	G: 14 cm^3

Appendix p. a30

A: 24 cm^3	B: 24 cm^3
C: 14 cm^3	D: 10 cm^3
E: 21 cm^3	F: 7 cm^3

Test

Tests, Unit 11, 4A and 4B, pp. 187-197

Enrichment

Use both multilink cubes (2 cm on a side) and 1-cm cubes. Put them next to each other so your student can compare the size. Ask her to measure the side of the multilink cube, and then to build a cube the same size using the 1-cm cubes. 8 cubes are needed. Ask her for the volume (8 cm^3). Have her build larger cubes or other shapes with the multilink cubes and find the volume in cubic centimeters. For example, if she builds a cube with 8 multilink cubes, the volume is 64 cm^3. Each multilink cube has a volume of 8 cm^3, so 8 of them have a volume of 64 cm^3.

Explore the relative sizes of volumes in other units:

Use the net on appendix p. a31. Cut it out and fold at the creases to make a cube. Ask your student to measure the side in inches and tell you the volume. The volume is 1 cubic inch, or 1 in.^{3}. A volume of 1 in.^{3} is as big as a cube with side 1 in.

Tape some paper together to make a similar net with 6 square feet and make a cube from it. Or use 12 strips of cardboard one foot long and use them to construct the edges of a cube. The volume is 1 cubic foot, or 1 ft^3. Have your student compare its volume visually with a cubic inch. Save these two paper cubes for an activity in the next chapter.

Use masking tape to mark a square with 1-meter sides on the floor with one side against the wall. Mark another square meter on the wall with the same edge as the square on the floor. Use meter sticks to mark two opposite vertical edges. Tell your student to imagine a cube from this. Its volume is 1 cubic meter, or 1 m^3. Have him compare its volume visually with a cubic centimeter.

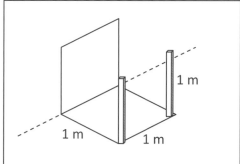

Chapter 5 – Volume of Rectangular Prisms

Objectives

- Find the volume of rectangular prisms, given the length, width, and height.
- Convert cm^3 to liters and milliliters.
- Solve problems involving the volume of liquids in rectangular containers.

Material

- Centimeter cubes
- Multilink cubes
- Centimeter graph paper
- 1000-cube from a base-10 set
- Liter measuring cup
- Dropper or teaspoon

Notes

In *Primary Mathematics* 3B, students learned to find the area of a rectangle given its length and width. In this chapter, your student will learn to find the volume of a rectangular prism by multiplying its length by its width by its height.

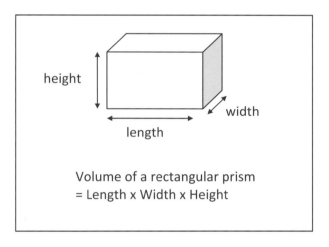

Volume of a rectangular prism
= Length x Width x Height

In *Primary Mathematics* 2 and *Primary Mathematics* 3, students learned to find the capacity of containers in liters and milliliters. In this chapter your student will learn that liters and milliliters are used to measure the volume of liquid in a container, not just the capacity of the whole container. Since she has used measuring cups or beakers to determine the capacity of other containers, this is not really a new concept. However, she will learn that 1 cm^3 is equivalent to 1 milliliter, and 1000 cm^3 is equivalent to 1 liter, and to convert cubic centimeters to liters and milliliters and vice versa.

Volume, not just capacity, is also measured in cups, quarts, and gallons, such as using a measuring cup to measure out a volume of liquid to add to a recipe, or how many gallons of gasoline to add to the car tank. As your student advances in science, the metric system will be used more often.

Although the liter is the more common unit for volume, the international standard unit of volume is the cubic meter, which is equal to 1000 liters.

(1) Find the volume of rectangular prisms

Activity

Use multilink cubes or unit cubes. Form a single layer rectangle. Ask your student for the shape of the solid. It is a rectangular prism. Ask him for the volume. For example, make a rectangle with the cubes that is 4 by 2. Point out that he can find the volume by multiplying the number of cubes along the length by the number of cubes along the width. Also point out that the height is 1 unit.

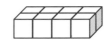

Number of cubes:
4 x 2 x 1 = 8

Add another layer and ask for the volume. Since we know how much is in one layer, we can find the number in both layers by multiplying the number of cubes on one layer by 2.

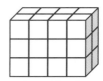

Number of cubes:
4 x 2 x 2 = 16

Add another layer and ask for the volume. There are now 3 layers with 4 x 2 in each layer.

The length is 4 units, the width is 2 units, and the height is 3 units. We can find the volume by multiplying these measurements together.

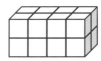

Number of cubes:
4 x 2 x 3 = 24

Point out that the order in which we multiply the sides does not matter. If we had a height of 5 cubes, we could multiply the height by the width first. 5 x 2 x 4 = 10 x 4 = 40. This is a little easier to calculate mentally than 4 x 2 x 5 = 8 x 5 = 40.

You can continue for other layers, or repeat with other examples.

Volume
= length x width x height
= 4 units x 2 units x 3 units
= 24 cubic units

Discussion

Concept p. 140

The volume is 24 cm^3.

This page shows pictorially the same thing as the previous concrete activity. The main concept that you want to be sure your student understands is that we don't have to see the unit cubes to be able to determine the volume. We can just be told the measurements for length, width, and height. By multiplying the units together, we get the number of unit cubes that would be used to make a rectangular prism with length, width, and height of 4 cm, 2 cm, and 3 cm respectively.

You can illustrate this concretely by making a box. Use centimeter graph paper to trace a box that is 9 cm by 8 cm. Use the pattern at the right. Cut along the solid lines and fold along the dotted lines, overlapping at the edges, to get an open box. Have your student measure the sides of the box and multiply length x width x height, and then count the number of unit cubes that will fit inside. The volume is the same using both methods.

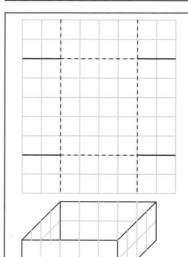

Volume = 5 cm x 4 cm x 2 cm
= 40 cm^3

Tasks 1-5, pp. 141-143

2, 3: Point out that these pictures are not actual size, they are drawn to scale. If you did not do the enrichment activity from the previous lesson, you may want to give your student a rough idea of the size of 1 cubic inch, 1 cubic foot, 1 cubic meter, and the size of the rectangular prism in Task 2. It will look narrower in "real life" than the picture in the textbook.

1. (a) 3 cm	(b) 5 cm	
2 cm	3 cm	
2 cm	1 cm	
12 cm^3	15 cm^3	
(c) 3 cm	(d) 4 cm	
3 cm	2 cm	
3 cm	5 cm	
27 cm^3	40 cm^3	

2. 80; 80 in.^{3}

3. 2; 2 m^3

4. 60 m^3

5. 27 m^3

Practice

Task 6, p. 143

6. A: 54 cm^3	B: 30,000 cm^3
C: 350 m^3	D: 180 m^3

Ask your student to find the volume of some rectangular prisms where she is only given the measurements and not a picture, such as those given below. See if she can solve the problems mentally by changing the order of the multiplication; for example, in the first problem 15 x 4 is 60, which is easy to multiply mentally by 12, whereas 15 x 12 is not as easy to multiply mentally.

⇒ Find the volume of a box with a length of 15 cm, a width of 12 cm, and a height of 4 cm.   (720 cm^3)

⇒ Find the volume of a box that is 11 cm by 2 cm by 8 cm.   (176 cm^3)

⇒ Find the volume of 5-centimeter cube.   (125 cm^3)

Workbook

Exercise 5, pp. 151-152 (answers p. 147)

Enrichment

Write down and then have your student solve the following problems.

⇒ A rectangular prism is made from 2-centimeter cubes. Its dimensions are 10 cubes by 8 cubes by 4 cubes. What is its volume?

Method 1: Number of cubes: 10 x 8 x 4 = 320
            Volume of 1 cube = 2 cm x 2 cm x 2 cm = 8 cm^3
            Total volume = 8 cm^3 x 320 = 2560 cm^3
Method 2: Dimensions: 10 x 2 cm = 20 cm, 8 x 2 cm = 16 cm, 4 x 2 cm = 8 cm
            Total volume: 20 cm x 16 cm x 8 cm = 2560 cm^3

⇒ A rectangular container is 11 cm long, 11 cm wide, and 9 cm high. How many 2-centimeter cubes can it hold?

Ask your student why we cannot simply find the total volume and divide by the volume of a 2-cm cube. If we put a row in the bottom layer, we will get a gap of 1 cm along both the length and width where we cannot fit a whole cube. So we need to first find how many cubes can fit along the length, width, and height. 11 ÷ 2 = 5 R 1; 9 ÷ 2 = 4 R 1.

Total number of cubes: 5 x 5 x 4 = 100

(2) Practice

Practice

Practice B, p. 145

You can do this practice now to consolidate the concepts from the last lesson, or after the next lesson. It is put in the guide now so that you can do the enrichment activity for your lesson, which will help with some concepts in the next lesson, and use the practice for independent work.

Enrichment

Draw a cube and label the sides of the cube as 1 ft. Ask your student to find the volume in cubic inches. Point out that just because there are 12 inches in a foot, we cannot say there are 12 cubic inches in a cubic foot. If you made the paper cubic foot and inch in the earlier enrichment activity, put the cubic inch inside the cubic foot. He should see that a lot more than 12 will fit. To find the volume of a cubic foot in cubic inches, we need to change the measurement units for each side to inches, and then use that to find the volume in cubic inches. There are 1728 cubic inches in a cubic foot.

Draw another cube or re-label the sides of the previous drawing with 1 yd. Ask students to find the volume in cubic feet. There are 27 cubic feet in a cubic yard.

Repeat with a cubic meter. There are 1 million cubic centimeters in a cubic meter.

---

1. (a) 9 in.^{3} (b) 15 in.^{3}

2. 30 cm x 25 cm x 15 cm = **11,250 cm^3**

3. 5 cm x 5 cm x 5 cm = **125 cm^3**

4. 12 ft x 10 ft x 3 ft = **360 ft^3**

5. 8 x 5 x 3 = **120 cubes**

6. 30 cm x 20 cm x 20 cm = **12,000 cm^3**

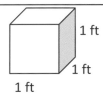

Volume of 1 ft^3 in in.^{3}:
1 ft x 1 ft x 1 ft = 1 ft^3
1 ft = 12 in.
12 in. x 12 in. x 12 in. = 1728 in.^{3}
1 ft^3 = 1728 in.^{3}

Volume of 1 yd^3 in ft^3:
1 yd x 1 yd x 1 yd = 1 yd^3
1 yd = 3 ft
3 ft x 3 ft x 3 ft = 27 ft^3
1 yd^3 = 27 ft^3

Volume of 1 m^3 in cm^3:
100 cm x 100 cm x 100 cm
= 1,000,000 cm^3

(3) Convert between cubic centimeters and liters

Activity

Have your student fill a measuring cup with 1 liter of water. If you have a dropper, use it to squeeze out about 20 drops. Tell your student that this is 1 milliliter. If you do not have a dropper, measure out a teaspoon of water into another container and tell her that this is about 5 milliliters. Ask her how many milliliters are in a liter. There are 1000 milliliters in a liter.

Show your student a centimeter cube or the unit cube from a base-10 set. Tell him that if we made a waterproof box that just fit around this cube, its capacity would be exactly 1 milliliter. 1 milliliter of water fills the same amount of space as 1 cubic centimeter.

Show your student the 1000-cube from a base-10 set. Ask her how many milliliters of water a container that fits exactly around the cubes would hold. Since 1 ml is the same as 1 cm^3, and there are 1000 cm^3 in a block, it would hold 1000 ml. Compare the cube to the liter of water in the measuring cup. Tell her that if we could pour a liter into the cube, it would fill it up completely. The volume of the cube is the same as 1 liter. 1 liter is 1000 cm^3.

Ask your student if a liter is the same as 1 m^3, since a liter is the same as 1000 cm^3. If you did the enrichment activities from the previous lessons, he would know it is not. If you did not do them, tell him that a cubic meter is 100 cm on the side (not 10, as in the thousand-cube), so its volume is 100 cm x 100 cm x 100 cm = 1,000,000 cm^3. So a cubic meter is a million cubic centimeters, or a thousand liters. We cannot say that 1000 cm^3 = 1 m^3 or that 1 m^3 = 1 liter.

1 ℓ = 1000 ml
1 ml = 1 cm^3
1 ℓ = 1000 cm^3
1000 cm^3 is NOT 1 m^3

Draw and label a box 20 cm by 10 cm by 12 cm. Ask your student to find the volume in cubic centimeters, in milliliters, and in liters and milliliters.

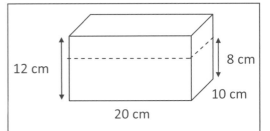

Volume = 20 cm x 10 cm x 12 cm
　　　　= 2400 cm^3
　　　　= 2400 ml
　　　　= 2 ℓ 400 ml

Tell your student that this box is a tank and is filled with water to 8 cm. Ask her for the volume of water in the tank, and how much more water is needed to fill the tank. Discuss two methods to find the answer to the second part; subtract from the total volume or multiply the length, width, and height of the remaining part of the tank.

Volume of water = 20 cm x 10 cm x 8 cm
　　　　　　　 = 1600 ml
　　　　　　　 = 1 ℓ 600 ml

Additional water needed to fill tank:
2 ℓ 400 ml – 1 ℓ 600 ml = 800 ml
　　　　　or
20 cm x 10 ml x 4 ml = 800 ml

Discussion

Tasks 7-10, p. 144

7: This task reiterates the concrete introduction in the previous activity.

8-9: Since 1 ml = 1 cm^3 and 1 ℓ = 1000 cm^3, the process for converting from liters to cubic centimeters is the same as converting from liters to milliliters and vice versa.

Practice

Practice C, p. 146

Workbook

Exercise 6, pp. 153-154 (answers p. 147)

Reinforcement

Extra Practice, Unit 11, Exercise 5, pp. 181-182

Test

Tests, Unit 11, 5A and 5B, pp. 199-208

---

7. 1000 cm^3
   1000 cm^3
   1 cm^3

8. (a) 2000 cm^3 (b) 400 cm^3 (c) 1200 cm^3

9. (a) 1 ℓ 750 ml   (b) 2 ℓ 450 ml   (c) 3 ℓ 50 ml

10. (a) 12,000 cm^3

   (b) 4800 cm^3
       4 ℓ 800 ml

---

1. (a) 3000 cm^3 (b) 250 cm^3 (c) 2060 cm^3

2. (a) 1 ℓ 50 ml   (b) 1 ℓ 800 ml   (c) 3 ℓ 500 ml

3. 15 cm x 10 cm x 3 cm = 450 cm^3 = **450 ml**
   The tin can hold 450 ml.

4. (a) 18 cm x 20 cm x 8 cm = **2880 cm^3**
   (b) 2880 cm^3 = 2880 ml = **2 ℓ 880 ml**

5. (a) 25 cm x 30 cm x 20 cm = **15,000 cm^3**
   (b) 15 cm x 10 cm x 20 cm = **3000 cm^3**

Review 11

Review

Review 11, pp. 147-152

Workbook

Review 11, pp. 155-163 (answers pp. 147-148)

Tests

Tests, Units 1-11 Cumulative Tests A and B, pp. 209-220

---

1 (a) 79,431; 79,433; 80,331; 80,431
  (b) 0.09, 0.55, 0.6, 0.7
  (c) $2\frac{2}{9}, 2\frac{4}{9}, 2\frac{2}{3}, \frac{9}{2}$
  (d) −5, -4, −3, 0, 1, 2

2. (a) $1\frac{1}{4}$ (b) $1\frac{2}{9}$ (c) $4\frac{7}{10}$
   (d) $\frac{1}{6}$ (e) 8   (f) 60

3. Length of flower bed: 25 m − 14 m = 11 m
   Width of flower bed: 20 m − 14 m = 6 m
   Area of flower bed: 11 m x 6 m = **66 m²**

4. (a) 340 ml
   (b) 744 g
   (c) 5 in.
   (d) 125 ml x 14 = 1750 ml = **1 ℓ 750 ml**
       He drinks 1 ℓ 750 ml in two weeks.

5. (a) 6 km 330 m   (b) 12 lb 6 oz
   (c) 4 h 10 min   (d) 33 ft 7 in.
   (e) 10 ℓ 350 ml   (f) 44 yd
   (g) 8 h 15 min   (h) 1 h 40 min

6. $\frac{3}{5}$ x 600 g = 3 x 120 = **360 g**

   She used 360 g of potatoes.

7. 6 x 2 x 3 = **36**
   36 cubes are needed.

8. (a) 2 units = $18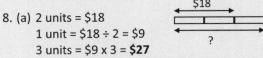
       1 unit = $18 ÷ 2 = $9
       3 units = $9 x 3 = **$27**
       The total amount of money is $27.
   (b) 8

9. (a) 1.6   (b) $2\frac{1}{20}$

10. (a) 22,548
    (b) 38,412
    (c) 14,455

11. (a) <   (b) <
    (c) $\frac{7}{10} < \frac{7}{9}$ (d) $3\frac{3}{6} > 2\frac{1}{6}$
    (e) 290 cm < 300 cm   (f) 18 qt > 6 qt

(Continued next page.)

12. (a) missing part: $n = 7009 - 1243 = \mathbf{5766}$
    (b) missing factor: $n = 332 \div 4 = \mathbf{83}$
    (c) missing factor: $n = 980 \div 4 = \mathbf{245}$
    (d) missing whole; $n = 217 \times 7 = \mathbf{1519}$
    (e) thousandths; $n = \mathbf{1000}$
    (f) $5 \times 9 = 45$, so $5 \times 0.09 = 0.45$; $n = \mathbf{0.09}$
    (g) $(4 + 3) \times n = 7 \times n = 70$; $n = \mathbf{10}$
    (h) move 2 from 132 to 140; $n = \mathbf{142}$
    (i) $36 \times 25 = 9 \times 4 \times 25 = 9 \times 100$; $n = \mathbf{100}$
    (j) $0.8 = \frac{8}{10} = \frac{4}{5}$; $n = \mathbf{4}$
    (k) $n = \mathbf{5}$
    (l) Rename 1 as $\frac{3}{3}$; $n = 3 + 2 = \mathbf{5}$

13. 9

14.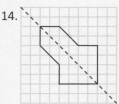

15.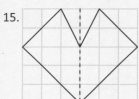

16.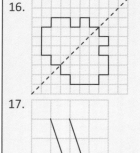

17.

18. (a) Perimeter: 110 cm
       Area: 460 cm^2
    (b) Perimeter: 60 m
       Area: 128 m^2

19. (a) Area of outside rectangle: 21 m x 11 m = 231 m^2
       Area of inside rectangle: 15 m x 5 m = 75 m^2
       Area of shaded part: 231 m^2 − 75 m^2 = **156 m^2**
    (b) Area of large rectangle: 13 m x 11 m = 143 m^2
       Area of small rectangle: 5 m x 7 m = 35 m^2
       Area of shaded part: 143 m^2 − 35 m^2 = **108 m^2**

20. Length: 78 m^2 ÷ 6 m = 13 m
    Perimeter: 13 m + 6 m + 13 m + 6 m = **38 m**

21. Side: 48 in. ÷ 4 = **12 in.**
    Area: 12 in. x 12 in. = **144 in.^{2}**

22. 4.5 m ÷ 5 = **0.9 m** or **90 cm**
    She used 0.9 m for each pillowcase.

23. 5 weeks: $70.50
    1 week: $70.50 ÷ 5 = $14.10
    8 weeks: $14.10 x 8 = **$112.80**
    She would save $112.80 in 8 weeks.

24.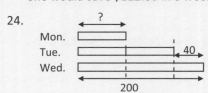

    2 units = 200 − 40 = 160
    1 unit = 160 ÷ 2 = **80**
    80 tickets were sold on Monday.

25. Find the number of possible outcomes.

    H ← R → Ham on rye
        W → Ham on white
        S → Ham on sourdough

    T ← R → Turkey on rye
        W → Turkey on white
        S → Turkey on sourdough

    B ← R → Beef on rye
        W → Beef on white
        S → Beef on sourdough

    P ← R → Pastrami on rye
        W → Pastrami on white
        S → Pastrami on sourdough

    He can choose **12** types of sandwiches.

26. (a) $\frac{3}{10}$
    (b) $\frac{1}{5}$

27. (a) Abe: $8
       Luigi: $10
    (b) March
    (c) March
    (d) Abe: $8 + $14 + $17 + $14 + $11 = $64
       Luigi: $10 + $14 + $7 + $19 + $15 = $65
       Luigi saved **$1** more.

Workbook

Exercise 4, p. 150

1. (a) 6 cm^3 (b) 6 cm^3
   (c) 18 cm^3 (d) 16 cm^3
   (e) 6 cm^3 (f) 9 cm^3

Exercise 5, pp. 151-152

1.

Solid	Length	Width	Height	Volume
B	2 in.	2 in.	2 in.	8 in.^{3}
C	5 in.	2 in.	4 in.	40 in.^{3}
D	3 in.	2 in.	7 in.	42 in.^{3}
E	7 in.	3 in.	2 in.	42 in.^{3}

2. 18 cm^3
   200 cm^3
   126 cm^3
   192 cm^3
   240 cm^3

Exercise 6, pp. 153-154

1. (a) 300 cm^3 (b) 800 cm^3

2. (a) 400 ml   (b) 120 ml

3. (a) 4 ℓ   (b) 3 ℓ

4. 1 ℓ 200 ml   3 ℓ 600 ml
   1 ℓ 200 ml   3 ℓ 600 ml
   2 ℓ 160 ml   1 ℓ 440 ml

Review 11, pp. 155-163

1. 2 km, 20 m, 253 cm, 2 m 35 cm

2. (a) 1000
   (b) 0.1

3. 40.6

4. (a) 6
   (b) 0

5. (a) 1, 2, 4, 5, 10, 20
   (b) 1, 2, 4
   (c) 2, 3, 5, 7

6. 3500 + 500 + 9600 = 13,600

7. 147.3 lb

8. 4.32

9. $\frac{3}{8}$

10. 1.5

11. (a) 2634 m
   (b) 5107 g
   (c) 184 min
   (d) 4 h 20 min
   (e) 4 kg 7 g
   (f) 5 m 80 cm
   (g) 3 ℓ 20 ml
   (h) 6 lb 12 oz

12. 250 ml x 6 = 1500 ml = **1 ℓ 500 ml**

13.
1 kg 680 g − 800 g = 880 g
880 g + 1 kg 680 g = **2 kg 560 g**

14. 6 ft − 1 ft 3 in. = 4 ft 9 in.
4 ft 9 in. − 1 ft 8 in. = **3 ft 1 in.**

15. 7 lb 8 oz ÷ 6 = 6 lb 24 oz ÷ 6 = **1 lb 4 oz**

16. 2 kg 450 g − 865 g = **1 kg 585 g**

17. Check drawing.

18. (a) 1 ℓ 540 ml
   (b) 3650 m

(Continued next page.)

Workbook

19. 1 ℓ 200ml ÷ 3 = 1200 ml ÷ 3 = **400 ml**

20. 3.82 m x 6 = **22.92 m**

21. 8 ft ÷ 6 = **1.3 ft**

22. Cost of shrimp = $1.50 x 5 = $7.50
Total cost = $7.50 + $4.50 = **$12**

23. Amount left: = $1.50 - $0.50 = $1.00
Fraction left = $\frac{100}{150}$ = $\frac{2}{3}$

Or, there are three 50 cents in $1.50, 1 out of 3
was used, so $\frac{2}{3}$ are left.

24. $\frac{2}{5}$ of $840 = 2 x $168 = **$336**

25. (45 yd + 20 yd) x 2 = 65 yd x 2 = 130 yd
130 yd x 5 = **650 yd**

26. 9 cm^3

27. 15 m x 6 m x 4 m = **360 m^3**

28. Width of card: 30 cm + 6 cm = 36 cm
Length of card: 24 cm + 6 cm = 30 cm
(36 cm + 30 cm) x 2 = 66 cm x 2 = **132 cm**
The perimeter of the card is 132 cm.

29.
```
┌───┬───┬───┬───┐
└───┴───┴───┴───┘
  ◄──────────►◄────►
     $45        ?
```
3 units = $45
1 unit = $45 ÷ 3 = **$15**
He had $15 left.

30. $\frac{1}{7}$ x 30.1 gal = **4.3 gal**

The bucket has a capacity of 4.3 gal.

31. Width = 1 unit, Length = 2 units,
Perimeter = 6 units = 30 in.
1 unit = 30 ÷ 6 = 5 in. = Width
2 units = 5 in. x 2 = 10 in. = Length
Area = 5 in. x 10 in. = **50 in.^{2}**
The area of the rectangle is 50 in.^{2}

32. 15 quarters = $0.25 x 15 = $3.75
35 nickels = $0.05 x 35 = $1.75
21 dimes = $0.10 x 21 = $2.10
$3.75 + $1.75 + $2.10 = **$7.60**
He saved $7.60 in the three months.

33. $1 has 4 quarters
$116 has $116 x 4 = 464 quarters
$0.75 has 3 quarters
464 + 3 = **467**
There are 467 quarters in $116.75.

34. Assume order is significant.

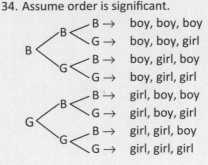

35. (a)

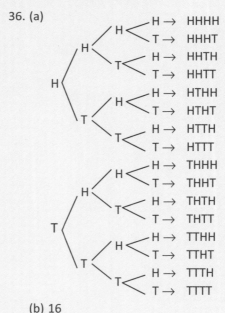

(b) 5 hours
(c) 12 hours
(d) 8 hours
(e) 7 or 8 hours (bimodal)

36. (a)

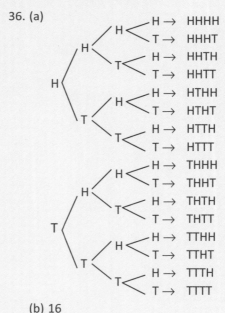

(b) 16

Mental Math 1	Mental Math 2	Mental Math 3		Mental Math 4	Mental Math 5	Mental Math 6
$\frac{1}{2} = $ **0.5**	0.25 + **0.75** = 1	0.34 + 0.03 = **0.37**		3.67 + 0.4 = **4.07**	0.003 + 0.002 = **0.005**	6.002 + 0.05 = **6.052**
$\frac{1}{5} = $ **0.2**	0.75 + **0.25** = 1	0.89 − 0.03 = **0.86**		6.88 − 0.06 = **6.82**	0.014 − 0.003 = **0.011**	1.506 − 0.3 = **1.206**
$\frac{3}{5} = $ **0.6**	0.99 + **0.01** = 1	0.45 + 0.4 = **0.85**		2.32 − 0.7 = **1.62**	0.126 + 0.01 = **0.136**	3.896 + 0.002 = **3.898**
$\frac{1}{20} = $ **0.05**	0.04 + **0.96** = 1	0.68 − 0.5 = **0.18**		1.6 − 0.08 = **1.52**	0.209 + 0.8 = **1.009**	0.119 + 0.03 = **0.149**
$\frac{7}{20} = $ **0.35**	0.52 + **0.48** = 1	0.38 + 0.3 = **0.68**		6.36 − 0.08 = **6.28**	2.231 − 0.007 = **2.224**	1.648 − 0.2 = **1.448**
$\frac{1}{4} = $ **0.25**	0.65 + **0.35** = 1	0.22 − 0.04 = **0.18**		1.64 + 0.07 = **1.71**	5.198 − 0.008 = **5.19**	7.12 − 0.009 = **7.111**
$\frac{3}{10} = $ **0.3**	0.33 + **0.67** = 1	0.72 + 0.5 = **1.22**		8.4 − 0.06 = **8.34**	6.218 + 0.05 = **6.268**	4.028 + 0.002 = **4.03**
$\frac{3}{50} = $ **0.06**	0.95 + **0.05** = 1	0.18 − 0.09 = **0.09**		2.28 + 0.8 = **3.08**	0.171 − 0.06 = **0.111**	8.1 + 0.005 = **8.105**
$\frac{2}{5} = $ **0.4**	0.42 + **0.58** = 1	0.73 + 0.9 = **1.63**		3.7 − 0.04 = **3.66**	9.01 + 0.009 = **9.019**	6.991 − 0.4 = **6.591**
$\frac{1}{25} = $ **0.04**	0.62 + **0.38** = 1	0.52 + 0.08 = **0.6**		5.19 + 0.01 = **5.2**	8.842 + 0.6 = **9.442**	0.4 − 0.004 = **0.396**
$\frac{3}{4} = $ **0.75**	0.55 + **0.45** = 1	0.93 − 0.2 = **0.73**		7.48 − 0.5 = **6.98**	0.142 − 0.04 = **0.102**	1.875 − 0.003 = **1.872**
$\frac{13}{100} = $ **0.13**	0.91 + **0.09** = 1	0.89 − 0.05 = **0.84**		6.44 + 0.07 = **6.51**	7.541 − 0.003 = **7.538**	4.172 − 0.8 = **3.372**
$2\frac{11}{25} = $ **2.44**	0.84 + **0.16** = 1	0.66 − 0.04 = **0.62**		1.32 + 0.8 = **2.12**	3.9 + 0.08 = **3.98**	7.052 + 0.007 = **7.059**
$1\frac{9}{20} = $ **1.45**	0.45 + **0.55** = 1	0.92 − 0.05 = **0.87**		9.99 + 0.9 = **10.89**	1.472 + 0.8 = **2.272**	9.204 − 0.2 = **9.004**
$\frac{3}{2} = $ **1.5**	0.07 + **0.93** = 1	0.22 + 0.05 = **0.27**		9.99 + 0.09 = **10.08**	2.355 + 0.005 = **2.36**	2.632 − 0.06 = **2.572**
	0.71 + **0.29** = 1	0.47 + 0.7 = **1.17**		1.5 − 0.03 = **1.47**	9.105 − 0.5 = **8.605**	3.311 + 0.7 = **4.011**
	0.62 + **0.38** = 1	0.17 + 0.08 = **0.25**		5 − 0.3 = **4.7**	3.4 + 0.03 = **3.43**	0.985 + 0.006 = **0.991**
	0.23 + **0.77** = 1	0.88 − 0.4 = **0.48**		5 − 0.03 = **4.97**	3.4 − 0.03 = **3.37**	6.324 − 0.09 = **6.234**
	0.19 + **0.81** = 1	0.9 + 0.06 = **0.96**		10 − 0.7 = **9.3**	3.4 − 0.003 = **3.397**	9.999 + 0.001 = **10**
	0.35 + **0.65** = 1	0.6 − 0.03 = **0.57**		10 − 0.07 = **9.93**	10 − 0.005 = **9.995**	9.999 + 0.01 = **10.009**

Mental Math 7	Mental Math 8	Mental Math 9		Mental Math 10	Mental Math 11	Mental Math 12
5.1 + 0.9 = **6**	0.72 + 0.06 = **0.78**	0.52 + 0.73 = **1.25**		4.9 − 0.5 = **4.4**	0.85 − 0.02 = **0.83**	0.98 − 0.03 = **0.95**
8.8 + 0.2 = **9**	0.48 + 0.6 = **1.08**	0.48 + 0.34 = **0.82**		9.6 − 0.3 = **9.3**	5.69 − 0.08 = **5.61**	1.84 − 0.04 = **1.8**
9.3 + 0.2 = **9.5**	0.09 + 0.59 = **0.68**	0.32 + 0.78 = **1.1**		4.2 − 0.8 = **3.4**	0.1 − 0.08 = **0.02**	4.73 − 0.09 = **4.64**
7.9 + 0.6 = **8.5**	0.63 + 0.5 = **1.13**	0.67 + 0.43 = **1.1**		3.3 − 0.7 = **2.6**	0.9 − 0.04 = **0.86**	3.5 − 0.02 = **3.48**
8.2 + 4.1 = **12.3**	0.16 + 0.04 = **0.2**	0.91 + 0.49 = **1.4**		2.5 − 0.6 = **1.9**	9.6 − 0.08 = **9.52**	2.3 − 0.07 = **2.23**
4.3 + 5.7 = **10**	0.62 + 0.8 = **1.42**	0.95 + 0.35 = **1.3**		8.1 − 0.9 = **7.2**	6.5 − 0.07 = **6.43**	4.66 − 0.09 = **4.57**
6.6 + 1.5 = **8.1**	0.92 + 0.08 = **1**	0.29 + 0.98 = **1.27**		7.3 − 0.6 = **6.7**	4.3 − 0.02 = **4.28**	3.53 − 0.05 = **3.48**
6.6 + 2.6 = **9.2**	0.42 + 0.8 = **1.22**	0.59 + 0.62 = **1.21**		5.4 − 0.8 = **4.6**	3.95 − 0.03 = **3.92**	2.6 − 0.06 = **2.54**
8 + 4.2 = **12.2**	0.91 + 0.03 = **0.94**	0.24 + 0.87 = **1.11**		3.7 − 0.2 = **3.5**	1 − 0.07 = **0.93**	2.42 − 0.08 = **2.34**
9.5 + 1.3 = **10.8**	0.58 + 0.6 = **1.18**	0.36 + 0.77 = **1.13**		6.5 − 0.2 = **6.3**	2 − 0.09 = **1.91**	4.55 − 0.09 = **4.46**
5.2 + 0.7 = **5.9**	0.62 + 0.09 = **0.71**	0.42 + 0.99 = **1.41**		6.3 − 0.7 = **5.6**	8 − 0.06 = **7.94**	2.6 − 0.04 = **2.56**
0.8 + 3.4 = **4.2**	0.86 + 0.04 = **0.9**	0.28 + 0.44 = **0.72**		8.2 − 0.3 = **7.9**	1 − 0.38 = **0.62**	6.92 − 0.09 = **6.83**
3.2 + 7.8 = **11**	0.65 + 0.8 = **1.45**	0.92 + 0.95 = **1.87**		9.4 − 0.6 = **8.8**	1 − 0.76 = **0.24**	1.22 − 0.05 = **1.17**
6.7 + 4.5 = **11.2**	0.34 + 0.06 = **0.4**	0.42 + 0.97 = **1.39**		8.5 − 0.5 = **8**	1 − 0.33 = **0.67**	4.85 − 0.06 = **4.79**
9.1 + 4.9 = **14**	0.03 + 0.69 = **0.72**	0.82 + 0.71 = **1.53**		7.9 − 0.4 = **7.5**	3 − 0.49 = **2.51**	2.32 − 0.06 = **2.26**
9.5 + 3.5 = **13**	0.41 + 0.9 = **1.31**	0.48 + 0.65 = **1.13**		6.21 − 0.9 = **5.31**	5 − 0.75 = **4.25**	2.83 − 0.08 = **2.75**
2.9 + 9.8 = **12.7**	0.78 + 0.07 = **0.85**	0.34 + 0.55 = **0.89**		4.36 − 0.9 = **3.46**	6 − 0.84 = **5.16**	0.36 − 0.08 = **0.28**
5.9 + 6.2 = **12.1**	0.49 + 0.02 = **0.51**	0.68 + 0.84 = **1.52**		8.04 − 0.6 = **7.44**	8 − 0.43 = **7.57**	3.74 − 0.07 = **3.67**
2.4 + 8.7 = **11.1**	4.48 + 0.9 = **5.38**	0.99 + 0.44 = **1.43**		3.24 − 0.7 = **2.54**	3 − 0.85 = **2.15**	2.87 − 0.08 = **2.79**
3.6 + 7.7 = **11.3**	3.27 + 0.7 = **3.97**	0.77 + 0.82 = **1.59**		2.3 − 0.4 = **1.9**	5 − 0.66 = **4.34**	2.43 − 0.06 = **2.37**

Answers to Mental Math

Mental Math 13	Mental Math 14	Mental Math 15
6.9 – 2.8 = **4.1**	5.54 – 0.98 = **4.56**	0.4 x 8 = **3.2**
5.6 – 3.2 = **2.4**	7.22 – 1.99 = **5.23**	0.7 x 7 = **4.9**
5.6 – 1.8 = **3.8**	3.83 – 2.95 = **0.88**	0.2 x 9 = **1.8**
9.7 – 6.4 = **3.3**	7.47 – 4.96 = **2.51**	0.06 x 2 = **0.12**
9.5 – 3.8 = **5.7**	8.09 – 3.98 = **4.11**	0.03 x 8 = **0.24**
5.2 – 4.8 = **0.4**	6.66 + 4.95 = **11.61**	7 x 0.5 = **3.5**
8.4 – 4.5 = **3.9**	3.97 + 2.22 = **6.19**	6 x 0.06 = **0.36**
6 – 3.7 = **2.3**	8.71 – 2.96 = **5.75**	0.09 x 8 = **0.72**
4.1 – 2.2 = **1.9**	6.02 – 4.98 = **1.04**	0.7 x 8 = **5.6**
5.8 – 2.3 = **3.5**	3.87 + 3.95 = **7.82**	0.3 x 9 = **2.7**
8 – 6.5 = **1.5**	7.01 – 2.98 = **4.03**	0.06 x 4 = **0.24**
4.2 – 2.4 = **1.8**	5.57 – 3.97 = **1.6**	4 x 0.05 = **0.2**
6.4 – 2.5 = **3.9**	1.99 + 6.23 = **8.22**	6 x 0.9 = **5.4**
5.5 – 2.8 = **2.7**	4.03 – 2.96 = **1.07**	0.8 x 2 = **1.6**
8.4 – 6.1 = **2.3**	5.2 – 3.99 = **1.21**	0.03 x 6 = **0.18**
5.3 – 3.8 = **1.5**	6.4 – 2.98 = **3.42**	0.5 x 3 = **1.5**
7 – 3.6 = **3.4**	3.7 + 0.98 = **4.68**	0.8 x 5 = **4**
7.7 – 3.8 = **3.9**	4.8 + 4.97 = **9.77**	0.06 x 7 = **0.42**
8.1 – 4.9 = **3.2**	3.05 + 0.98 = **4.03**	9 x 0.09 = **0.81**
5.3 – 3.5 = **1.8**	6.5 – 0.95 = **5.55**	0.2 x 5 = **1**

Mental Math 16		Mental Math 17
7.7 x 3 = 21 + 2.1 = **23.1**		1.5 ÷ 5 = **0.3**
3.4 x 5 = 15 + **2** = **17**		6.4 ÷ 8 = **0.8**
8.3 x 6 = 48 + 1.8 = **49.8**		0.42 ÷ 7 = **0.06**
4.3 x 4 = 16 + 1.2 = **17.2**		0.3 ÷ 6 = **0.05**
4.5 x 6 = 24 + **3** = **27**		5.4 ÷ 9 = **0.6**
8.5 x 5 = 40 + 2.5 = **42.5**		0.24 ÷ 4 = **0.06**
6.2 x 7 = 42 + 1.4 = **43.4**		0.2 ÷ 5 = **0.04**
3.3 x 5 = 15 + 1.5 = **16.5**		8.1 ÷ 9 = **0.9**
0.43 x 3 = 1.2 + 0.09 = **1.29**		0.49 ÷ 7 = **0.07**
0.63 x 8 = 4.8 + 0.24 = **5.04**		4 ÷ 8 = **0.5**
2.3 x 2 = **4.6**	2.8 x 3 = **8.4**	0.27 ÷ 9 = **0.03**
5.6 x 3 = **16.8**	8.2 x 7 = **57.4**	0.21 ÷ 3 = **0.07**
3.6 x 7 = **25.2**	4.9 x 4 = **19.6**	0.25 ÷ 5 = **0.05**
7.1 x 2 = **14.2**	3.6 x 2 = **7.2**	4.5 ÷ 5 = **0.9**
2.1 x 4 = **8.4**	0.14 x 2 = **0.28**	0.16 ÷ 2 = **0.08**
6.4 x 5 = **32**	0.26 x 3 = **0.78**	3.2 ÷ 4 = **0.8**
4.7 x 3 = **14.1**	0.83 x 3 = **2.49**	3.5 ÷ 5 = **0.7**
3.1 x 4 = **12.4**	0.27 x 5 = **1.35**	2.4 ÷ 8 = **0.3**
6.9 x 8 = **55.2**	0.55 x 4 = **2.2**	0.48 ÷ 6 = **0.08**
		1.8 ÷ 6 = **0.3**

Mental Math 18	Mental Math 19
6.26 + 0.4 = **6.66**	10 min 5 s – 50 s = **9** min **15** s
0.06 x 5 = **0.3**	15 km 5 m – 50 m = **14** km **955** m
0.174 – 0.01 = **0.164**	20 m 5 cm – 50 cm = **19** m **55** cm
7.8 + 2.5 = **10.3**	8 ft 5 in – 10 in. = **7** ft **7** in.
6.36 – 0.08 = **6.28**	13 lb 5 oz – 10 oz = **12** lb **11** oz
0.63 ÷ 7 = **0.09**	11 min 50 s + 50 s = **12** min **40** s
0.1 – 0.06 = **0.04**	21 km 600 m + 600 m = **22** km **200** m
6.3 – 0.98 = **5.32**	5 m 60 cm + 60 cm = **6** m **20** cm
9.4 – 0.3 = **9.1**	3 ft 10 in + 10 in. = **4** ft **8** in.
5 – 0.04 = **4.96**	13 lb 10 oz + 10 oz = **14** lb **4** oz
0.04 x 8 = **0.32**	10 min 10 s x 10 = **101** min **40** s
6.73 – 0.08 = **6.65**	10 ℓ 10 ml x 10 = **100** ℓ **100** ml
4.56 + 0.99 = **5.55**	10 m 10 cm x 10 = **101** m **0** cm
2.6 – 0.07 = **2.53**	10 ft 10 in x 10 = **108** ft **4** in.
7.21 – 3.97 = **3.24**	10 lb 10 oz x 10 = **106** lb **4** oz
0.56 + 0.08 = **0.64**	10 gal 3 qt x 10 = **107** gal **2** qt
2.6 – 0.4 = **2.2**	10 yr 3 months x 10 = **102** yr **6** months
6.218 + 0.1 = **6.318**	10 weeks 3 days x 10 = **104** weeks **2** days
0.3 ÷ 5 = **0.06**	10 yd 1 ft x 10 = **103** yd **1** ft
1.5 – 0.8 = **0.7**	2 qt 1 pt x 10 = **25** qt **0** pt

Mental Math 1	Mental Math 2	Mental Math 3
$\dfrac{1}{2} =$ _____	$0.25 +$ _____ $= 1$	$0.34 + 0.03 =$ _____
$\dfrac{1}{5} =$ _____	$0.75 +$ _____ $= 1$	$0.89 - 0.03 =$ _____
$\dfrac{3}{5} =$ _____	$0.99 +$ _____ $= 1$	$0.45 + 0.4 =$ _____
$\dfrac{1}{20} =$ _____	$0.04 +$ _____ $= 1$	$0.\,68 - 0.5 =$ _____
$\dfrac{7}{20} =$ _____	$0.52 +$ _____ $= 1$	$0.38 + 0.3 =$ _____
$\dfrac{1}{4} =$ _____	$0.65 +$ _____ $= 1$	$0.22 - 0.04 =$ _____
$\dfrac{3}{10} =$ _____	$0.33 +$ _____ $= 1$	$0.72 + 0.5 =$ _____
$\dfrac{3}{50} =$ _____	$0.95 +$ _____ $= 1$	$0.18 - 0.09 =$ _____
$\dfrac{2}{5} =$ _____	$0.42 +$ _____ $= 1$	$0.73 + 0.9 =$ _____
$\dfrac{1}{25} =$ _____	$0.62 +$ _____ $= 1$	$0.52 + 0.08 =$ _____
$\dfrac{3}{4} =$ _____	$0.55 +$ _____ $= 1$	$0.93 - 0.2 =$ _____
$\dfrac{13}{100} =$ _____	$0.91 +$ _____ $= 1$	$0.89 - 0.05 =$ _____
$2\dfrac{11}{25} =$ _____	$0.84 +$ _____ $= 1$	$0.66 - 0.04 =$ _____
$1\dfrac{9}{20} =$ _____	$0.45 +$ _____ $= 1$	$0.92 - 0.05 =$ _____
$\dfrac{3}{2} =$ _____	$0.07 +$ _____ $= 1$	$0.22 + 0.05 =$ _____
	$0.71 +$ _____ $= 1$	$0.47 + 0.7 =$ _____
	$0.62 +$ _____ $= 1$	$0.17 + 0.08 =$ _____
	$0.23 +$ _____ $= 1$	$0.88 - 0.4 =$ _____
	$0.19 +$ _____ $= 1$	$0.9 + 0.06 =$ _____
	$0.35 +$ _____ $= 1$	$0.6 - 0.03 =$ _____

Mental Math 4	Mental Math 5	Mental Math 6
3.67 + 0.4 = _____	0.003 + 0.002 = _____	6.002 + 0.05 = _____
6.88 – 0.06 = _____	0.014 – 0.003 = _____	1.506 – 0.3 = _____
2.32 – 0.7 = _____	0.126 + 0.01 = _____	3.896 + 0.002 = _____
1.6 – 0.08 = _____	0.209 + 0.8 = _____	0.119 + 0.03 = _____
6.36 – 0.08 = _____	2.231 – 0.007 = _____	1.648 – 0.2 = _____
1.64 + 0.07 = _____	5.198 – 0.008 = _____	7.12 – 0.009 = _____
8.4 – 0.06 = _____	6.218 + 0.05 = _____	4.028 + 0.002 = _____
2.28 + 0.8 = _____	0.171 – 0.06 = _____	8.1 + 0.005 = _____
3.7 – 0.04 = _____	9.01 + 0.009 = _____	6.991 – 0.4 = _____
5.19 + 0.01 = _____	8.842 + 0.6 = _____	0.4 – 0.004 = _____
7.48 – 0.5 = _____	0.142 – 0.04 = _____	1.875 – 0.003 = _____
6.44 + 0.07 = _____	7.541 – 0.003 = _____	4.172 – 0.8 = _____
1.32 + 0.8 = _____	3.9 + 0.08 = _____	7.052 + 0.007 = _____
9.99 + 0.9 = _____	1.472 + 0.8 = _____	9.204 – 0.2 = _____
9.99 + 0.09 = _____	2.355 + 0.005 = _____	2.632 – 0.06 = _____
1.5 – 0.03 = _____	9.105 – 0.5 = _____	3.311 + 0.7 = _____
5 – 0.3 = _____	3.4 + 0.03 = _____	0.985 + 0.006 = _____
5 – 0.03 = _____	3.4 – 0.03 = _____	6.324 – 0.09 = _____
10 – 0.7 = _____	3.4 – 0.003 = _____	9.999 + 0.001 = _____
10 – 0.07 = _____	10 – 0.005 = _____	9.999 + 0.01 = _____

Mental Math 7	Mental Math 8	Mental Math 9
5.1 + 0.9 = _____	0.72 + 0.06 = _____	0.52 + 0.73 = _____
8.8 + 0.2 = _____	0.48 + 0.6 = _____	0.48 + 0.34 = _____
9.3 + 0.2 = _____	0.09 + 0.59 = _____	0.32 + 0.78 = _____
7.9 + 0.6 = _____	0.63 + 0.5 = _____	0.67 + 0.43 = _____
8.2 + 4.1 = _____	0.16 + 0.04 = _____	0.91 + 0.49 = _____
4.3 + 5.7 = _____	0.62 + 0.8 = _____	0.95 + 0.35 = _____
6.6 + 1.5 = _____	0.92 + 0.08 = _____	0.29 + 0.98 = _____
6.6 + 2.6 = _____	0.42 + 0.8 = _____	0.59 + 0.62 = _____
8 + 4.2 = _____	0.91 + 0.03 = _____	0.24 + 0.87 = _____
9.5 + 1.3 = _____	0.58 + 0.6 = _____	0.36 + 0.77 = _____
5.2 + 0.7 = _____	0.62 + 0.09 = _____	0.42 + 0.99 = _____
0.8 + 3.4 = _____	0.86 + 0.04 = _____	0.28 + 0.44 = _____
3.2 + 7.8 = _____	0.65 + 0.8 = _____	0.92 + 0.95 = _____
6.7 + 4.5 = _____	0.34 + 0.06 = _____	0.42 + 0.97 = _____
9.1 + 4.9 = _____	0.03 + 0.69 = _____	0.82 + 0.71 = _____
9.5 + 3.5 = _____	0.41 + 0.9 = _____	0.48 + 0.65 = _____
2.9 + 9.8 = _____	0.78 + 0.07 = _____	0.34 + 0.55 = _____
5.9 + 6.2 = _____	0.49 + 0.02 = _____	0.68 + 0.84 = _____
2.4 + 8.7 = _____	4.48 + 0.9 = _____	0.99 + 0.44 = _____
3.6 + 7.7 = _____	3.27 + 0.7 = _____	0.77 + 0.82 = _____

Mental Math 10	Mental Math 11	Mental Math 12
4.9 – 0.5 = _____	0.85 – 0.02 = _____	0.98 – 0.03 = _____
9.6 – 0.3 = _____	5.69 – 0.08 = _____	1.84 – 0.04 = _____
4.2 – 0.8 = _____	0.1 – 0.08 = _____	4.73 – 0.09 = _____
3.3 – 0.7 = _____	0.9 – 0.04 = _____	3.5 – 0.02 = _____
2.5 – 0.6 = _____	9.6 – 0.08 = _____	2.3 – 0.07 = _____
8.1 – 0.9 = _____	6.5 – 0.07 = _____	4.66 – 0.09 = _____
7.3 – 0.6 = _____	4.3 – 0.02 = _____	3.53 – 0.05 = _____
5.4 – 0.8 = _____	3.95 – 0.03 = _____	2.6 – 0.06 = _____
3.7 – 0.2 = _____	1 – 0.07 = _____	2.42 – 0.08 = _____
6.5 – 0.2 = _____	2 – 0.09 = _____	4.55 – 0.09 = _____
6.3 – 0.7 = _____	8 – 0.06 = _____	2.6 – 0.04 = _____
8.2 – 0.3 = _____	1 – 0.38 = _____	6.92 – 0.09 = _____
9.4 – 0.6 = _____	1 – 0.76 = _____	1.22 – 0.05 = _____
8.5 – 0.5 = _____	1 – 0.33 = _____	4.85 – 0.06 = _____
7.9 – 0.4 = _____	3 – 0.49 = _____	2.32 – 0.06 = _____
6.21 – 0.9 = _____	5 – 0.75 = _____	2.83 – 0.08 = _____
4.36 – 0.9 = _____	6 – 0.84 = _____	0.36 – 0.08 = _____
8.04 – 0.6 = _____	8 – 0.43 = _____	3.74 – 0.07 = _____
3.24 – 0.7 = _____	3 – 0.85 = _____	2.87 – 0.08 = _____
2.3 – 0.4 = _____	5 – 0.66 = _____	2.43 – 0.06 = _____

Mental Math 13	Mental Math 14	Mental Math 15
6.9 − 2.8 = _____	5.54 − 0.98 = _____	0.4 x 8 = _____
5.6 − 3.2 = _____	7.22 − 1.99 = _____	0.7 x 7 = _____
5.6 − 1.8 = _____	3.83 − 2.95 = _____	0.2 x 9 = _____
9.7 − 6.4 = _____	7.47 − 4.96 = _____	0.06 x 2 = _____
9.5 − 3.8 = _____	8.09 − 3.98 = _____	0.03 x 8 = _____
5.2 − 4.8 = _____	6.66 + 4.95 = _____	7 x 0.5 = _____
8.4 − 4.5 = _____	3.97 + 2.22 = _____	6 x 0.06 = _____
6 − 3.7 = _____	8.71 − 2.96 = _____	0.09 x 8 = _____
4.1 − 2.2 = _____	6.02 − 4.98 = _____	0.7 x 8 = _____
5.8 − 2.3 = _____	3.87 + 3.95 = _____	0.3 x 9 = _____
8 − 6.5 = _____	7.01 − 2.98 = _____	0.06 x 4 = _____
4.2 − 2.4 = _____	5.57 − 3.97 = _____	4 x 0.05 = _____
6.4 − 2.5 = _____	1.99 + 6.23 = _____	6 x 0.9 = _____
5.5 − 2.8 = _____	4.03 − 2.96 = _____	0.8 x 2 = _____
8.4 − 6.1 = _____	5.2 − 3.99 = _____	0.03 x 6 = _____
5.3 − 3.8 = _____	6.4 − 2.98 = _____	0.5 x 3 = _____
7 − 3.6 = _____	3.7 + 0.98 = _____	0.8 x 5 = _____
7.7 − 3.8 = _____	4.8 + 4.97 = _____	0.06 x 7 = _____
8.1 − 4.9 = _____	3.05 + 0.98 = _____	9 x 0.09 = _____
5.3 − 3.5 = _____	6.5 − 0.95 = _____	0.2 x 5 = _____

Mental Math 16	Mental Math 17

Mental Math 16

7.7 x 3 = 21 + 2.1 = _____

3.4 x 5 = _____ + _____ = _____

8.3 x 6 = _____ + _____ = _____

4.3 x 4 = _____ + _____ = _____

4.5 x 6 = _____ + _____ = _____

8.5 x 5 = _____ + _____ = _____

6.2 x 7 = _____ + _____ = _____

3.3 x 5 = _____ + _____ = _____

0.43 x 3 = _____ + _____ = _____

0.63 x 8 = _____ + _____ = _____

2.3 x 2 = _____	2.8 x 3 = _____
5.6 x 3 = _____	8.2 x 7 = _____
3.6 x 7 = _____	4.9 x 4 = _____
7.1 x 2 = _____	3.6 x 2 = _____
2.1 x 4 = _____	0.14 x 2 = _____
6.4 x 5 = _____	0.26 x 3 = _____
4.7 x 3 = _____	0.83 x 3 = _____
3.1 x 4 = _____	0.27 x 5 = _____
6.9 x 8 = _____	0.55 x 4 = _____

Mental Math 17

1.5 ÷ 5 = _____

6.4 ÷ 8 = _____

0.42 ÷ 7 = _____

0.3 ÷ 6 = _____

5.4 ÷ 9 = _____

0.24 ÷ 4 = _____

0.2 ÷ 5 = _____

8.1 ÷ 9 = _____

0.49 ÷ 7 = _____

4 ÷ 8 = _____

0.27 ÷ 9 = _____

0.21 ÷ 3 = _____

0.25 ÷ 5 = _____

4.5 ÷ 5 = _____

0.16 ÷ 2 = _____

3.2 ÷ 4 = _____

3.5 ÷ 5 = _____

2.4 ÷ 8 = _____

0.48 ÷ 6 = _____

1.8 ÷ 6 = _____

Mental Math 18	Mental Math 19
6.26 + 0.4 = _____	10 min 5 s – 50 s = _____ min _____ s
0.06 x 5 = _____	15 km 5 m – 50 m = _____ km _____ m
0.174 – 0.01 = _____	20 m 5 cm – 50 cm = _____ m _____ cm
7.8 + 2.5 = _____	8 ft 5 in – 10 in. = _____ ft _____ in.
6.36 – 0.08 = _____	13 lb 5 oz – 10 oz = _____ lb _____ oz
0.63 ÷ 7 = _____	11 min 50 s + 50 s = _____ min _____ s
0.1 – 0.06 = _____	21 km 600 m + 600 m = _____ km _____ m
6.3 – 0.98 = _____	5 m 60 cm + 60 cm = _____ m _____ cm
9.4 – 0.3 = _____	3 ft 10 in + 10 in. = _____ ft _____ in.
5 – 0.04 = _____	13 lb 10 oz + 10 oz = _____ lb _____ oz
0.04 x 8 = _____	10 min 10 s x 10 = _____ min _____ s
6.73 – 0.08 = _____	10 ℓ 10 ml x 10 = _____ ℓ _____ m
4.56 + 0.99 = _____	10 m 10 cm x 10 = _____ m _____ cm
2.6 – 0.07 = _____	10 ft 10 in x 10 = _____ ft _____ in.
7.21 – 3.97 = _____	10 lb 10 oz x 10 = _____ lb _____ oz
0.56 + 0.08 = _____	10 gal 3 qt x 10 = _____ gal _____ qt
2.6 – 0.4 = _____	10 yr 3 months x 10 = _____ yr _____ months
6.218 + 0.1 = _____	10 weeks 3 days x 10 = _____ weeks _____ days
0.3 ÷ 5 = _____	10 yd 1 ft x 10 = _____ yd _____ ft
1.5 – 0.8 = _____	2 qt 1 pt x 10 = _____ qt _____ pt

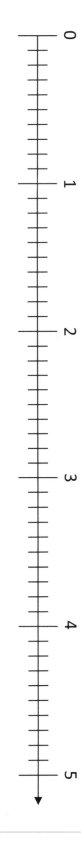

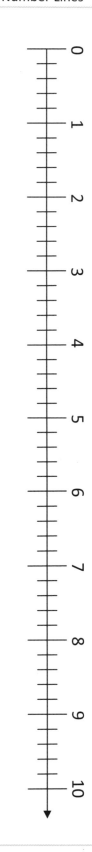

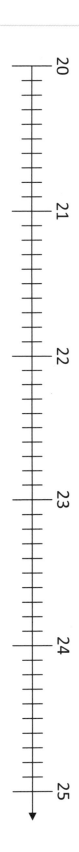

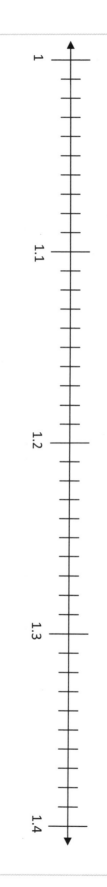

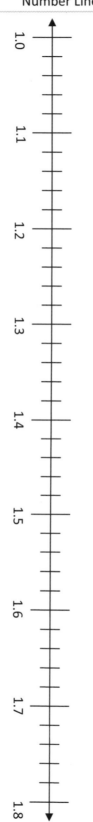

Square Dot Paper

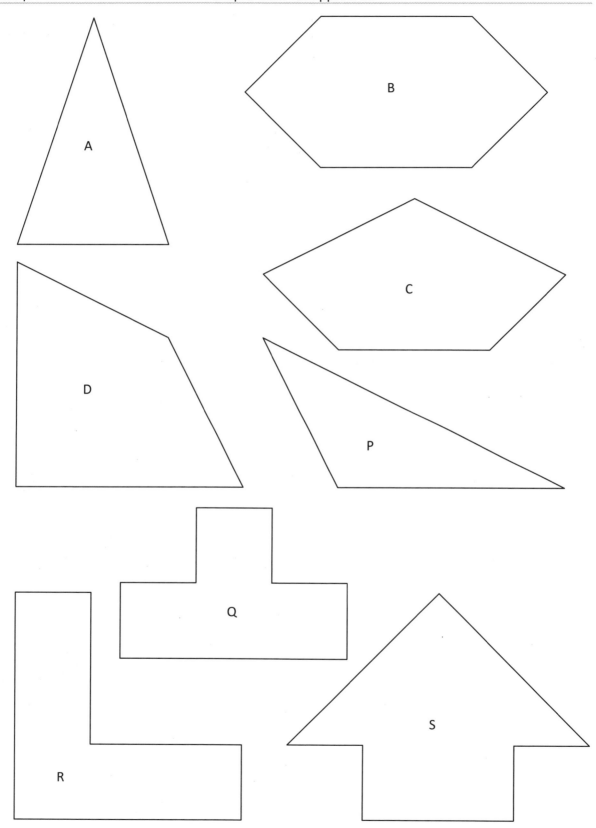

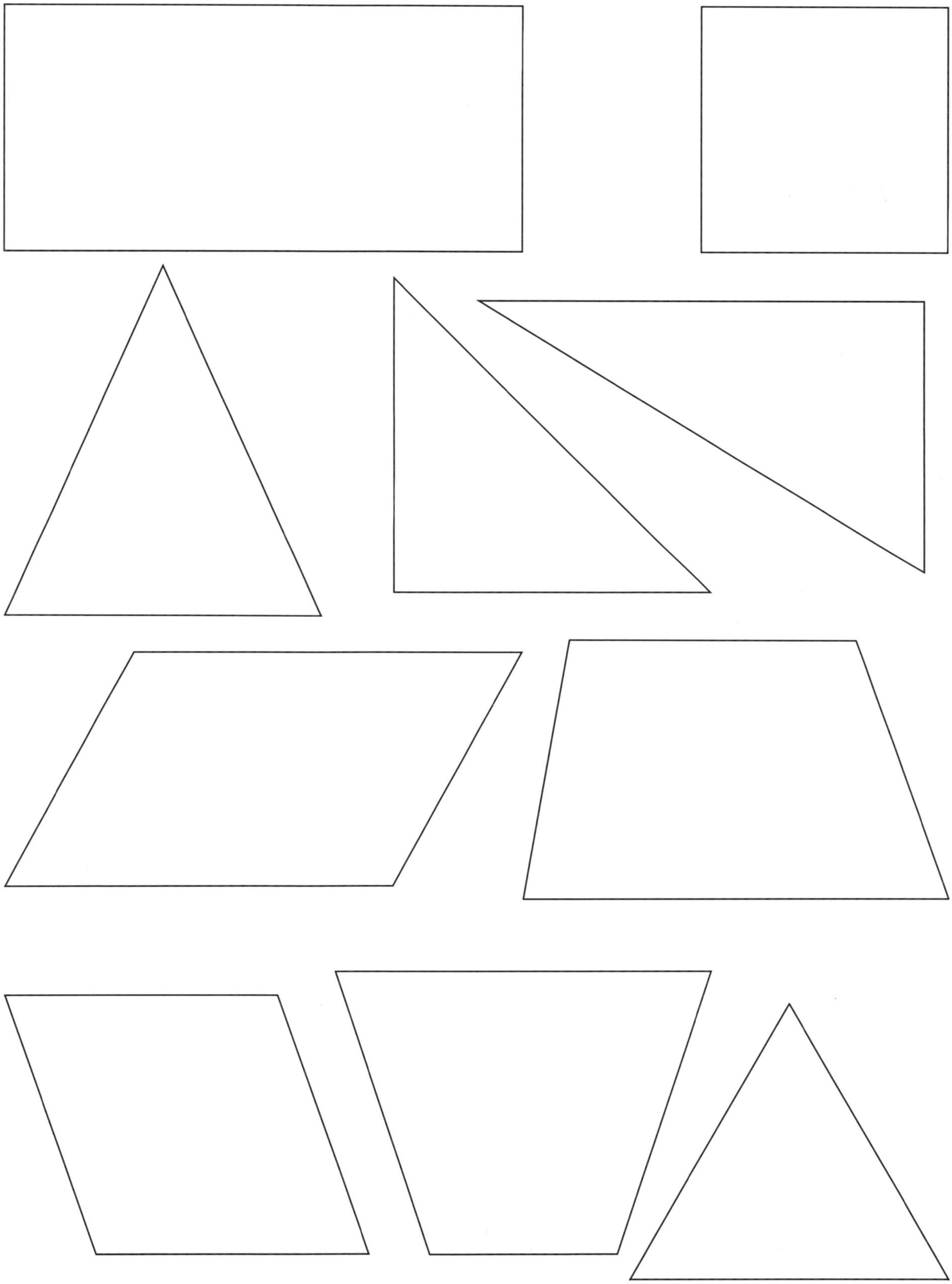

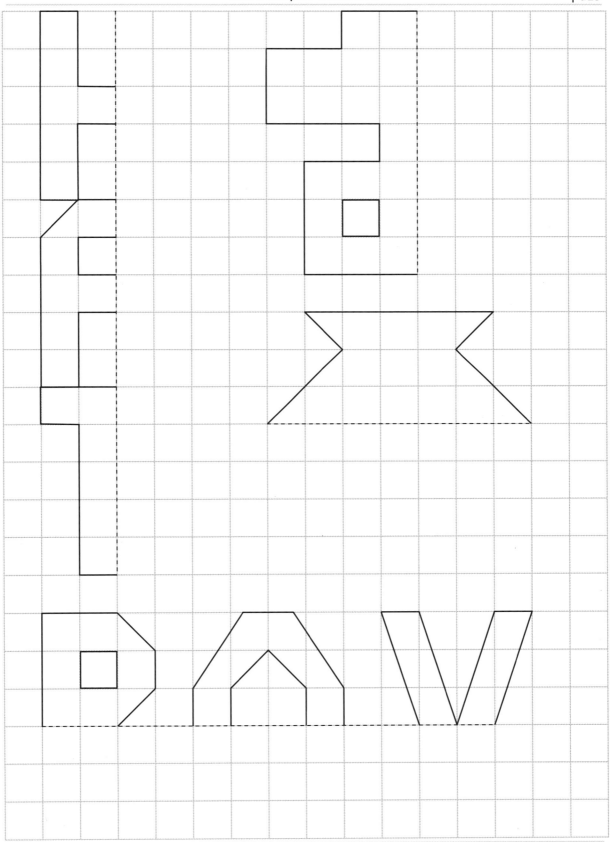

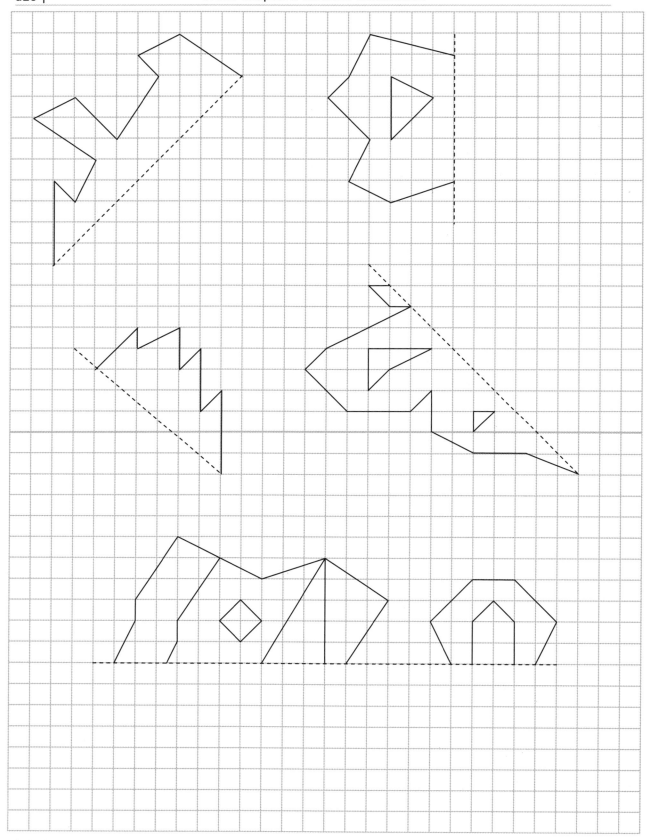

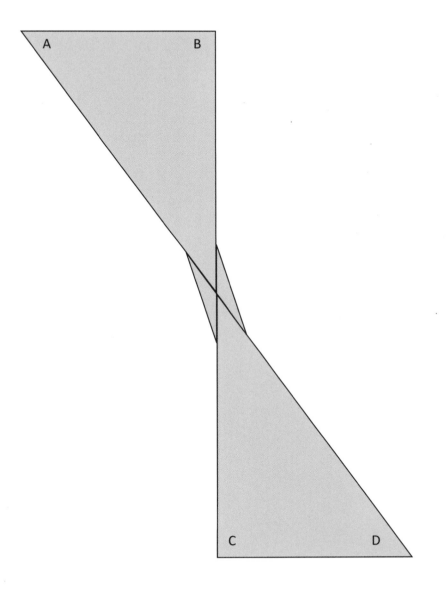

Which of the following figures have line symmetry, rotational symmetry, or no symmetry?

(a)

(b)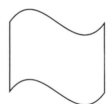

(c)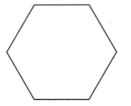

(d)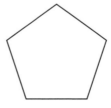

(e)

(f)

(g)

(h)

(i)

(j)

(k)

(l)

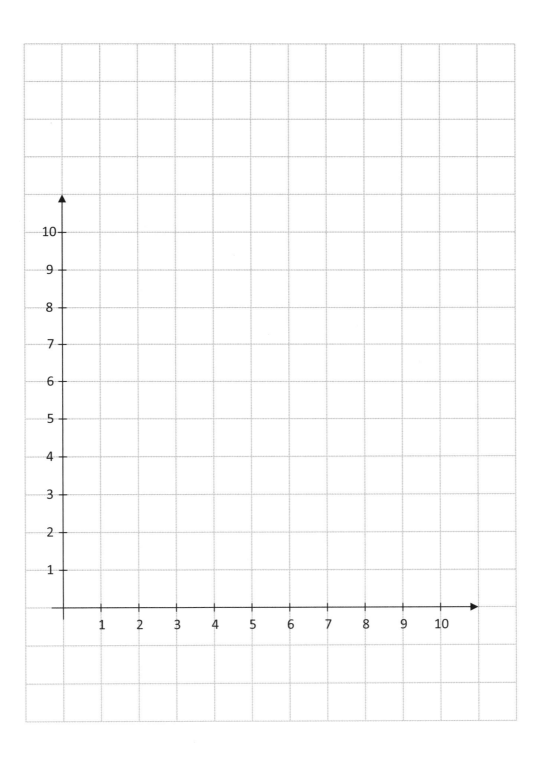

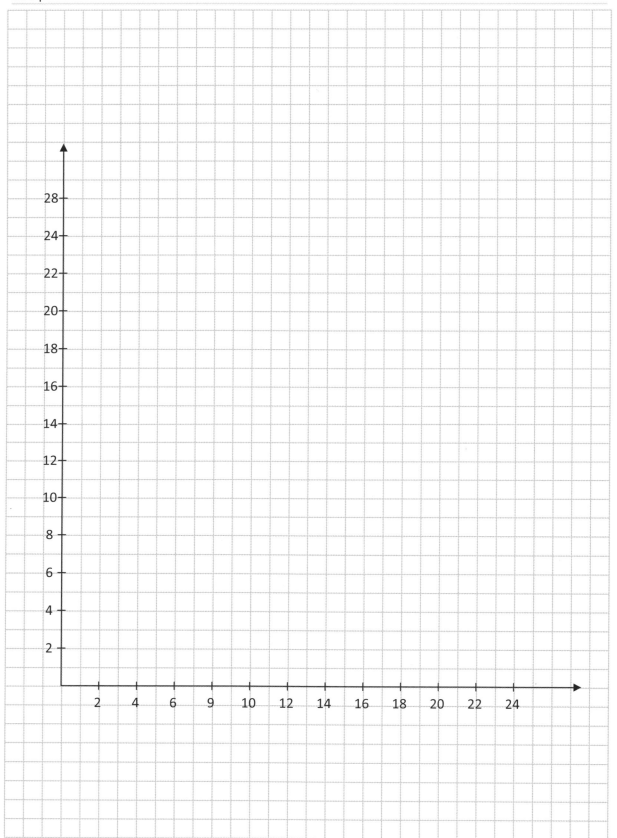

1st toss	2nd toss	3rd toss	4th toss

H
HH
HHH
HHHH
HHHT
HHT
HHTH
HHTT
HT
HTH
HTHH
HTHT
HTT
HTTH
HTTH
T
TH
THH
THHH
THHT
THT
THTH
THTT
TT
TTH
TTHH
TTHT
TTT
TTTH
TTTT

	$\dfrac{}{64} =$
	$\dfrac{}{64} =$
	$\dfrac{}{64} =$
	$\dfrac{}{64} =$
	$\dfrac{}{64} =$
	$\dfrac{}{64} =$
	$\dfrac{}{64} =$
	$\dfrac{}{64} =$
	$\dfrac{}{64} =$
	$\dfrac{}{64} =$
	$\dfrac{}{64} =$
	$\dfrac{}{64} =$
	$\dfrac{}{64} =$
	$\dfrac{}{64} =$
	$\dfrac{}{64} =$
	$\dfrac{}{64} =$

1 coin   2 coins   2 coins   4 coins

4 H, 0 T $\dfrac{}{64} =$

3 H, 0 T $\dfrac{}{64} =$

2 H, 0 T   3 H, 1 T $\dfrac{}{64} =$

1 H, 0 T   2 H, 1 T $\dfrac{}{64} =$

1 H, 1 T   2 H, 2 T $\dfrac{}{64} =$

0 H, 1 T   1 H, 2 T $\dfrac{}{64} =$

0 H, 2 T   1 H, 3 T $\dfrac{}{64} =$

0 H, 3 T $\dfrac{}{64} =$

0 H, 4 T $\dfrac{}{64} =$

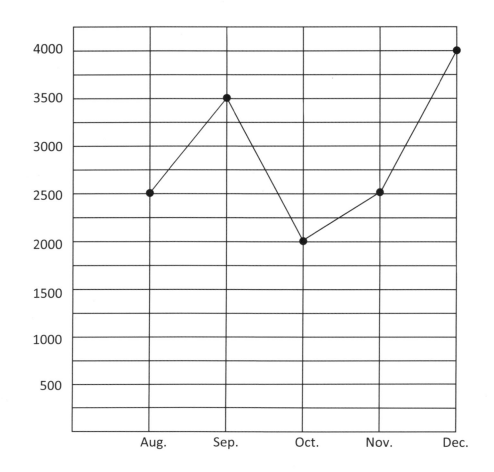

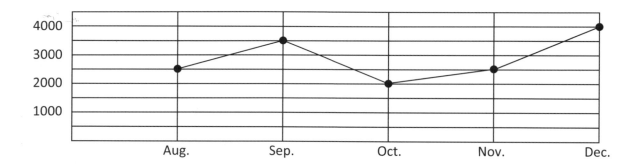

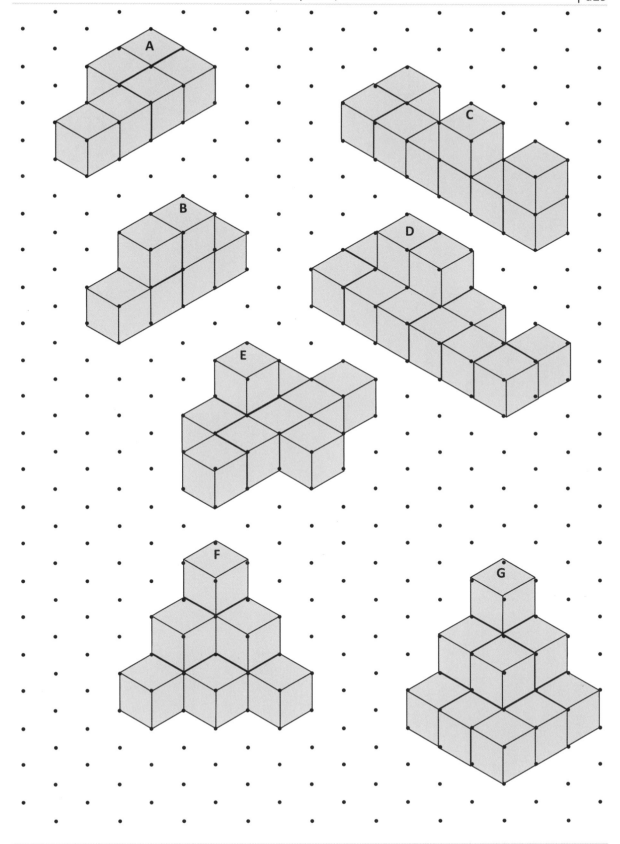

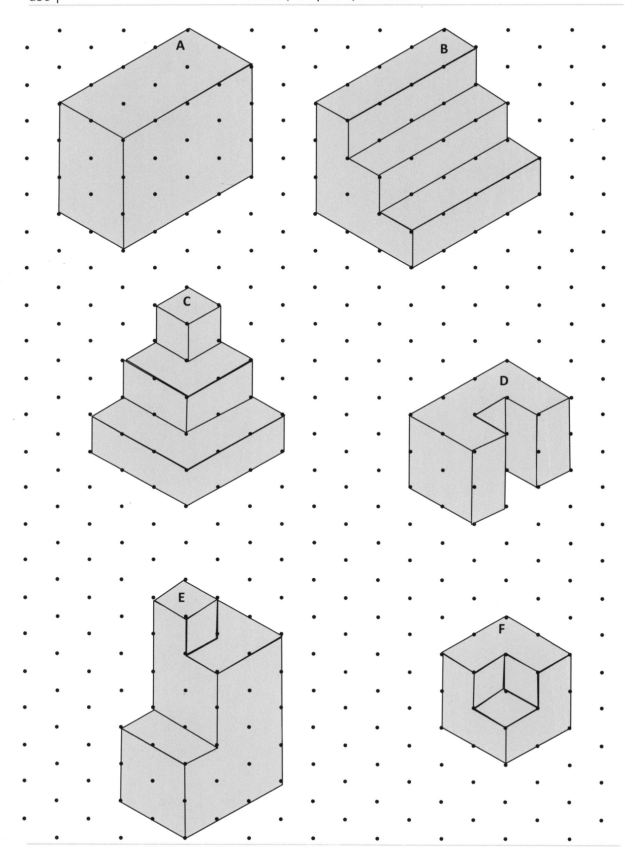

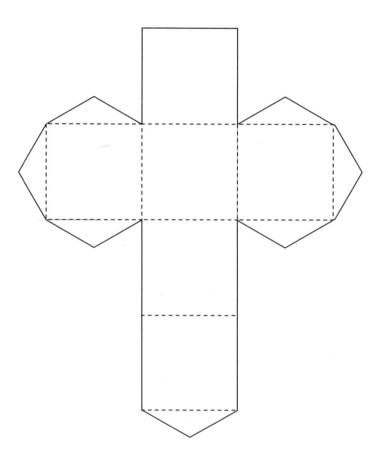

Blank Page